Study (

for use with

Basic Statistics for Business and Economics

Fifth Edition

Douglas A. Lind
Coastal Carolina University

William G. Marchal
University of Toledo

Samuel A. Wathen
Coastal Carolina University

Prepared by
Walter Lange

Boston Burr Ridge, IL Dubuque, IA Madison, WI New York San Francisco St. Louis Bangkok Bogotá Caracas Kuala Lumpur Lisbon London Madrid Mexico City Milan Montreal New Delhi Santiago Seoul Singapore Sydney Taipei Toronto

Study Guide for use with
BASIC STATISTICS FOR BUSINESS AND ECONOMICS
Douglas A. Lind/William G. Marchal/Samuel A. Wathen

Published by McGraw-Hill/Irwin, an imprint of The McGraw-Hill Companies, Inc., 1221 Avenue of the Americas, New York, NY 10020.

1 2 3 4 5 6 7 8 9 0 QPD/QPD 0 9 8 7 6 5 4

ISBN 0-07-298398-1

www.mhhe.com

The McGraw-Hill Companies

PREFACE

This study guide is especially designed to accompany the Fifth Edition of *Basic Statistics for Business and Economics* by Douglas A. Lind, William G. Marchal, and Samuel A. Wathen It can also be used alone, or as a companion to most other introductory statistics texts. It provides a valuable source of reinforcement for the material in the text. The chapters in the text and the study guide are parallel in topics, notation, and the numbering of formulas. Students will attain the most benefit if they study the textbook first, and then read the corresponding chapter in the study guide. The major features of the study guide include:

- **Chapter Goals**. They are listed first and stress the main concepts covered and the tasks students should be able to perform after having studied the chapter. It is recommended that students refer to the goals before reading the chapter to get an overview of the material to be studied and again after completing the chapter to confirm mastery of the material.

- **Brief Introduction.** A brief Introduction follows the goals. In capsule form the material covered in previous chapters is tied with that covered in the current chapter, thus maintaining continuity throughout the book.

- **Definitions**. Key words are defined and used in their correct statistical context.

Key Words are in a text box for easy reference.

- **Formulas.** The formulas are placed in a shaded formula box for easy reference.

Formula box is used to emphasize formulas

- A **glossary** follows the chapter discussion. The glossary provides definitions of the key words used in the chapter and is a handy reference.

- **Chapter Examples.** Chapter examples, including solutions, come next. In this section the step-by-step method of solution is presented along with an interpretation of the results. The values are kept small to emphasize the concept.

- **Self-Reviews.** Following the chapter examples is a self-review. The student completes the self-review and checks the answer in the Self-Review Answer Section at the end of the guide. Thus the student can check his/her comprehension of the material as he/she progresses through the chapter.

- **Chapter Assignments.** Chapter assignments cover the entire chapter and are intended to be completed outside the classroom. Part I of the assignment consists of multiple-choice questions, Part II consists of problems, with space for students to show essential work and a box for the answers. The pages are perforated, so that assignments can be torn out and handed in to the instructor for grading.

For this revised edition I wish to thank Dennis J. Heban for his invaluable assistance in reviewing the manuscript and checking the accuracy of the solutions to the examples, self-reviews, and assignments. A special thanks to Danuta T. Lange who prepared the camera-ready copy for this publication.

Walter H. Lange

TABLE OF CONTENTS

CHAPTER 1
WHAT IS STATISTICS?

Chapter Goals

After completing this chapter, you will be able to:

1. Understand why we study statistics.
2. Explain what is meant by ***descriptive statistics*** and ***inferential statistics***.
3. Distinguish between a ***qualitative variable*** and a ***quantitative variable***.
4. Distinguish between a ***discrete variable*** and a ***continuous variable***.
5. Distinguish among the ***nominal***, ***ordinal***, ***interval***, and ***ratio*** levels of measurement.
6. Define the terms ***mutually exclusive*** and ***exhaustive***.

Introduction

No doubt you have noticed the large number of facts and figures, often referred to as ***statistics,*** that appear in the newspapers and magazines you read, websites you visit, television you watch (especially sporting events), and in grocery stores where you shop. A simple figure is called a ***statistic*** (singular). A few examples:

- The newest data show high-income taxpayers earning more money; top 1% earned almost 1/5 of the nation's income and paid over 34% of the nation's federal individual income tax. (www.taxfoundation.org/prtopincome.html)
- Forest Laboratories, Inc. had a 25% increase in earnings per share for the first quarter of fiscal year 2005. (***Forest Laboratories News***, July 2004).
- The average sale price per square foot for multi-family sales of apartment buildings in 2003 was $32.43. (***Noneman Real Estate Investment Letter***, July 2004)
- During the week of July 31, 2004 North American car and truck production included 92,283 U.S. cars and 144,812 U.S. trucks. (www.autonews.com).
- The median price of an existing family home in Austin, Texas is $157,750, down 1.0% (***Austin Board of Realtor***, June 23, 2004).
- The Dow Jones Industrial Average (DJIA) closed this week at 10,139.71. (***Toledo Blade,*** July 30, 2004)

Forest Laboratories increase in earnings of 25 percent is a statistic (singular). The Dow Jones average of 10,139.71 is a statistic. A collection of figures is called statistics (plural). An example from the August 2, 2004 ***USA Today*** daily feature on the "Moneyline" is shown on the right.

Markets	
Dow Jones Industrial Average	10,139.71
NASDAQ composite	1887.36
T-bond, 30 year yield	5.20%
Oil, light sweet crude, barrel	$43.80
U.S. dollar, yen per dollar	111.22

You may think of statistics simply as a collection of numerical information. However, ***statistics*** has a much broader meaning.

> ***Statistics:*** The science of collecting, organizing, presenting, analyzing, and interpreting data to assist in making more effective decisions.

Note in this definition of statistics that the initial step is the collection of pertinent information. This information may come from newspapers or magazines, various websites, a company's human relations director, the local, state, or federal government, universities, nonprofit organizations, the United Nations, and so on. A few actual publications of the federal government and others are:

- ***Statistical Abstract of the United States***, published annually by the U.S. Department of Commerce.
- ***Monthly Labor Review***, published monthly by the U.S. Department of Labor.
- ***Survey of Current Business***, published monthly by the U.S. Department of Commerce.
- ***Social Security Bulletin***, published annually by the U.S. Social Security Administration.
- ***Crime in the United States***, published annually by the U.S. Federal Bureau of Investigation.
- ***Hospital Statistics***, published annually by the American Hospital Association.
- ***Vital Statistics of the United States***, published annually by the National Center for Health Statistics.

If the information is not available from company records or public sources, it may be necessary to conduct a ***survey.*** For example, the A.C. Nielsen Company surveys about 1,200 homes on an ongoing basis to determine which TV programs are being watched, and Gallup surveys registered voters before an election to estimate the percent that will vote for a certain candidate. These firms also sample the population regarding food preference, what features in automobiles are desirable, and what appliances consumers will most likely purchase next year.

Fortune magazine annually surveys 4,000 senior executives, outside directors, and securities analysts to evaluate the companies in their industry to find the ten most admired firms, and the least admired firms. Each executive is asked to rate a list of firms on eight attributes; namely, innovativeness, quality of management, quality of products and services, long-term investment value, financial soundness, employee talent, social responsibility to the community and the environment, and wise use of corporate assets. Each attribute is rated on a scale of zero (poor) to ten (excellent). The ten most admired companies are listed in the table.

Rank	Company
1	Wal-Mart Stores
2	Berkshire Hathaway
3	Southwest Airlines
4	General Electric
5	Dell Computer
6	Microsoft
7	Johnson & Johnson
8	Starbucks
9	Federal Express
10	IBM

Source: **Fortune**, February 23, 2004

Why Study Statistics?

Statistics is required for many college programs for three reasons.

1. **Numerical information is everywhere.** If you look in various newspapers (***USA Today, Wall Street Journal***), magazines (***Time, Business Week, Sports Illustrated, People***) you will be bombarded with numerical information. You need to be able to determine if the conclusions as reported are reasonable. Was the sample large enough? You must be able to read and interpret the charts or graphs.

2. **Statistical techniques are used to make decisions that affect our lives.** Insurance companies use statistics to determine the premiums you pay for automobile insurance, the Environmental Protection Agency uses various statistical tools to determine air quality in your area, and the Internal Revenue Service uses statistical surveys to determine if your tax return should be subject to an audit.

3. **Knowledge of statistical methods will help you understand why decisions are made and give you a better understanding of how they affect you.** No matter what line of work you select, you will find yourself faced with decisions where an understanding of data anaylsis is helpful.

Types of Statistics - Descriptive and Inferential Statistics

The definition of statistics referred to collecting, organizing, and presenting numerical information. Data stored in a computer's memory or in a filing cabinet are of little value. Techniques are available that organize this information in a more meaningful form. Such aids are called ***descriptive statistics***.

Descriptive statistics: Methods of organizing, summarizing, and presenting data in an informative way.

We often present statistical information in a graphical form. A statistical tool designed to describe the movement of a series of numbers over a long period of time (such as production, imports, wages and stock market trends) is called a line chart. The line chart below, for example, depicts the upward movement of the Dow Jones average of 30 industrials year-end closing prices since 1986.

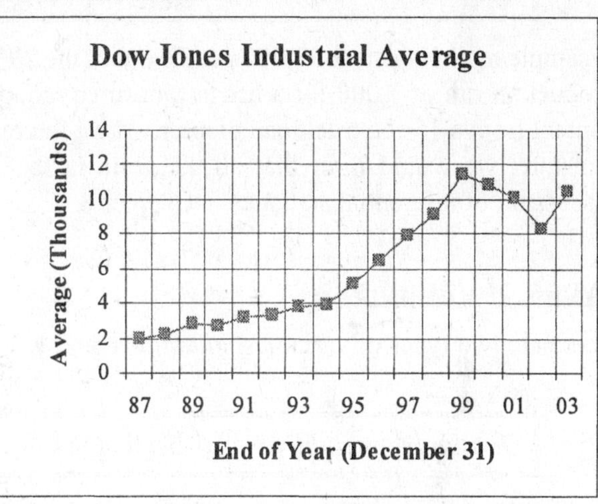

Notice how easy it is to describe the trend of stock prices: The price of the 30 industrials, as represented by the Dow, rose somewhat steadily from about 1,980 in 1986 to over 11,000 in 1999. The DJIA has shown a decline from the peak in 1999 through 2002 with an increase in 2003.

Another descriptive measure is referred to as an average. Some examples are:

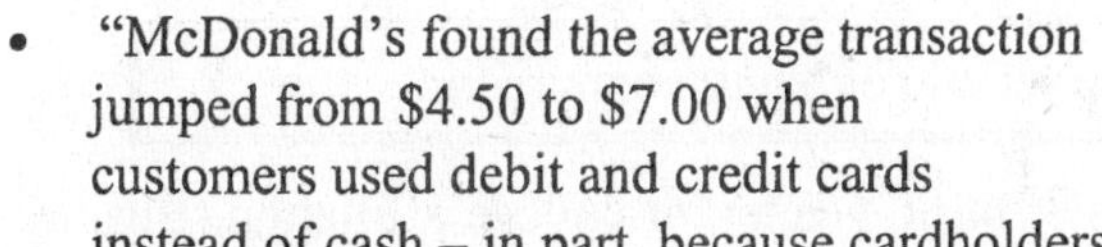

- "McDonald's found the average transaction jumped from $4.50 to $7.00 when customers used debit and credit cards instead of cash – in part, because cardholders tend to buy for more people". ***(Wall Street Journal,*** July, 23, 2004)

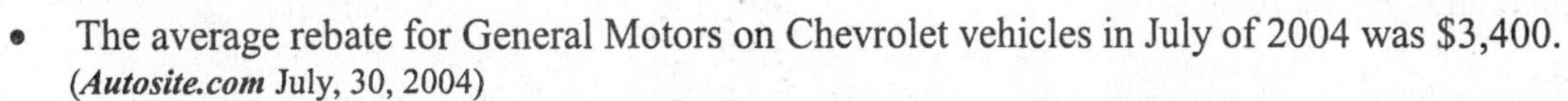

- The average rebate for General Motors on Chevrolet vehicles in July of 2004 was $3,400. (***Autosite.com*** July, 30, 2004)
- The 2003 National Corn Growers yield contest was won by the Schmeke brothers. They averaged 296.94 bushels of corn per acre. (www.ncga.com/03profits/cyc/main/index.htm)
- The median (an average) selling price of multi-family apartments in Lucas County, Ohio in 2003 was $148,000. (***Noneman Real Estate Investment Letter***, July 2004)

The Bureau of Labor Statistics describing the labor force in the United States reported that the average number of employed persons in April of 2004 was 146,741,000 and the average hourly earnings were $15.65. The average number of unemployed people was 8,164,000. Averages and other descriptive measures are presented in Chapter 3.

A second aspect of statistics is called ***inferential statistics***.

Inferential statistics: The methods used to determine something about a population, on the basis of a sample.

This branch of statistics deals with problems requiring the use of a sample to infer something about the ***population***.

> ***Population***: The entire set of individuals or objects of interest or the measurements obtained from all individuals or objects of interest.

A population might consist of all the 8,816,268 people in North Carolina, or all 494,423 people in Wyoming. Or, the population might consist of all the teams in the Canadian Football League, the Price/Earnings ratios for all chemical stocks, or the total assets of the 20 largest banks in the United States. A population, therefore, can be considered the total collection of people, prices, ages, square footage of homes being constructed in Flint, Michigan in 2004, and so on.

To infer something about a population, we usually take a ***sample*** from the population.

> ***Sample***: A portion, or part, of the population of interest.

A sample might consist of 2,000 people out of the 20,290,743 people in Texas, 12 headlights selected from a production run of 1,000 for a life test, or three scoops of grain selected at random to be tested for moisture content from a 15-ton truckload of grain. If we found that the three scoops of grain consisted of 9.50 percent moisture, we would infer that all the grain in the 15-ton load had 9.50 percent moisture. We start our discussion of inferential statistics in Chapter 8.

Types of Variables

There are two types of variables, ***qualitative*** and ***quantitative***.

> ***Qualitative variable***: A variable that has the characteristic of being nonnumeric.

A classification of students at your university by the state of birth, gender, or college affiliation (Business, Education, Liberal Arts, etc.) is an example of a qualitative variable. Other examples include: brand of soft drink, eye color, and type of vehicle.

> ***Quantitative variable***: A variable being studied that can be reported numerically.

Examples of quantitative variables include: the balance in your checking account, the ages of the members of the United States Congress, the speeds of automobiles traveling along I-70 in Kansas, the number of customers served yesterday at Bo-Rics in the Colonial Mall or the number of students enrolled in your statistics class.

There are two types of quantitative variables, ***discrete*** and ***continuous***.

> ***Discrete variable***: A quantitative variable that can only assume certain values. There is usually a "gap" between the values.

Examples of discrete variables are: the number of children in a family, the number of customers in a carpet store in an hour, or the number of commercials aired last hour on radio station WEND. A family can have two or three children, but not 2.445, or WEND can air five or six commercials, but not 5.75. Usually discrete variables result from counting.

> ***Continuous variable:*** A quantitative variable that can assume any value within a range.

Examples of continuous variables are: the amount of snow for the winter of 2003-2004 in Toronto, Ontario, the pressure in a tire, or a person's weight. Typically, continuous variables are the result of measuring something. We can measure the pressure in a tire, or the amount of snow in Toronto.

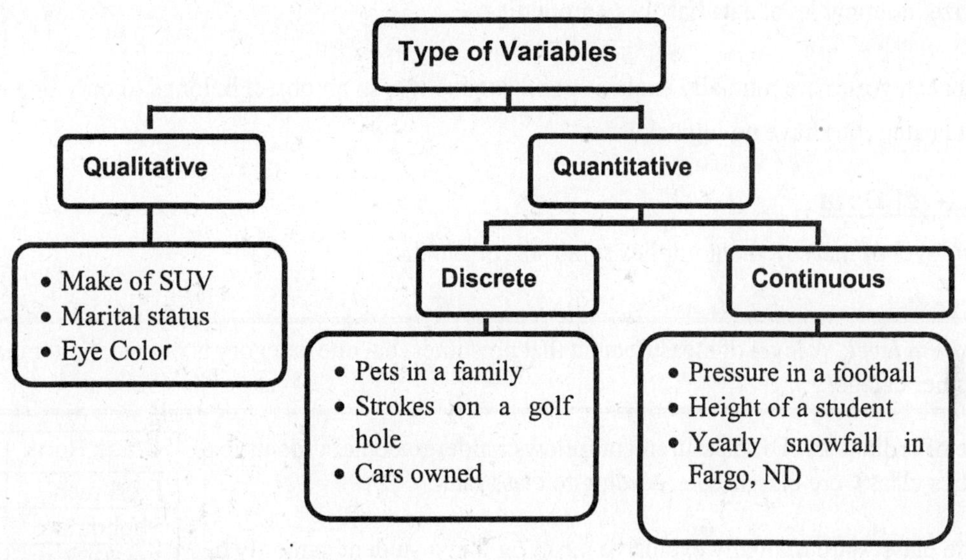

Levels of Measurement

Data may be classified into four classes or levels of measurement. These classes are nominal, ordinal, interval, and ratio. The level of measurement of the data often dictates the calculations that can be done to summarize and present the data.

Nominal Level Data

The nominal level of measurement is the lowest or most primitive level of measurement. When data can only be classified into classes, we refer to it as being the ***nominal*** level of measurement.

> ***Nominal level***: A level of measurement in which the data are sorted into classes with no particular order to the classes.

At this level the classes have no particular order or rank and are ***mutually exclusive.***

> ***Mutually exclusive***: A property of a set of categories such that an individual or object is included in only one category.

For example, www.automotivenews.com gave these counts for car and truck production in North American for the week of July 31, 2004.

North American car & truck production	Number Wk of 07/31/04
Cars	92,283
Trucks	144,812
Total	237,095

The data is nominal level of measurement because it can only be classified into classes (car or truck) and it is immaterial what order they are listed. Trucks could be listed first followed by cars. The classes are mutually exclusive meaning that the type of vehicle can be counted into only one category. Such categories are said to be ***exhaustive***.

> ***Exhaustive:*** A property of a set of categories such that each individual, object, or measurement must appear in a category.

This means that a vehicle must appear in one of the categories, it cannot be in both.

To summarize, nominal level data has these properties:

- Data categories are mutually exclusive and exhaustive, so an object belongs to only one category.
- Data categories have no logical order.

Ordinal Level Data

The ***ordinal*** level of measurement implies some sort of ranking.

> ***Ordinal level***: A level of measurement that presumes that one category is ranked higher than another category.

An example of ordinal level of measurement follows: undergraduate students in a basic statistics class were classified according to class rank.

Class Rank	Number
Freshman	13
Sophomore	17
Junior	9
Senior	5

Note that the classes are mutually exclusive meaning that a student can only be counted in one category. A student cannot be a freshman and a sophomore at the same time. Also, the classes are exhaustive meaning that a student must appear in one of the classes. In addition, a ranking of students is implied meaning that juniors are ranked "higher" than sophomores.

To summarize, ordinal level data has these properties:

- Data classifications are mutually exclusive and exhaustive.
- Data classifications have some logical order.
- Data classifications are ranked or ordered according to the particular trait they possess.

Interval Level Data

The ***interval level*** of measurement is the next highest level.

> ***Interval level:*** Includes the ranking characteristics of the ordinal scale and, in addition, the difference between values is a constant size.

Temperature on the Fahrenheit scale is an example. Suppose the high temperature for the last three days was 85, 73, and 78 degrees Fahrenheit. We can easily put the readings in a rank order, but in addition we can study the difference between readings. Why is this so? One degree on the Fahrenheit temperature scale is a constant unit of measurement. Note in this example that the zero point is just another point on the scale. It does not represent the absence of temperature, just that it is cold! Test scores are another example of the interval scale of measurement.

In addition to the constant difference characteristic, interval scaled data have all the features of nominal and ordinal measurements. Temperatures are mutually exclusive, that is, the high temperature yesterday cannot be both 88 and 85 degrees. The "greater than" feature of ordinal data permits the ranking of daily high temperatures.

The properties of the interval scale are:

- Data classifications are mutually exclusive and exhaustive.
- Data classifications are ordered according to the amount of the characteristic they possess.
- Equal differences in the characteristic are represented by equal differences in the measurement.

Ratio Level Data

The ***ratio level*** of measurement is the highest level of measurement.

> ***Ratio level***: Has all the characteristics of the interval scale, but additionally there is a meaningful zero point and the ratio between two values is meaningful.

Weight, height, and money are examples of the ratio scale of measurement. If you have $20 and your friend has $10, then you have twice as much money as your friend. The zero point represents the absence of money. That is, the zero point is fixed and represents the absence of the characteristic being measured. If you have zero dollars, you have none of the characteristic being measured.

The properties of the ratio level are:

- Data classifications are mutually exclusive and exhaustive.
- Data classifications are ordered according to the amount of the characteristic they posses.
- Equal differences in the characteristic are represented by equal differences in the numbers assigned to the classifications.
- The zero point reflects the absence of the characteristic.

Levels of Data			
Nominal Data	**Ordinal Data**	**Interval Data**	**Ratio Data**
↓	↓	↓	↓
Data may only be classified	Data are ranked	Meaningful difference between values	Meaningful 0 point and ratio between values
↓	↓	↓	↓
Hair color, zip code, make of truck	Order of finish, military rank, class rank	Score on test temperature, shoe size	Income, weight, distance traveled

Glossary

Continuous variable: A quantitative variable that can assume any value within a range.

Descriptive statistics: Methods of organizing, summarizing, and presenting data in an informative way.

Discrete variable: A quantitative variable that can only assume certain values. There is usually a "gap" between the values.

Exhaustive: A property of a set of categories such that each individual, object, or measurement must appear in a category.

Inferential statistics: The methods used to determine something about a population, on the basis of a sample.

Interval level: Includes the ranking characteristics of the ordinal scale and, in addition, the difference between values is a constant size.

Mutually exclusive: A property of a set of categories such that an individual or object is included in only one category.

Nominal level: A level of measurement in which the data are sorted into classes with no particular order to the classes.

Ordinal level: A level of measurement that presumes that one category is ranked higher than another category.

Population: The entire set of individuals or objects of interest or the measurements obtained from all individuals or objects of interest.

Qualitative variable: A variable that has the characteristic of being nonnumeric.

Quantitative variable: A variable being studied that can be reported numerically.

Ratio level: Has all the characteristics of the interval scale, but additionally there is a meaningful zero point and the ratio of two values is meaningful.

Sample: A portion, or part, of the population of interest.

Statistics: The science of collecting, organizing, presenting, analyzing, and interpreting data to assist in making more effective decisions.

CHAPTER 1 ASSIGNMENT

WHAT IS STATISTICS?

Name ______________________ Section __________ Score __________

Part I Classify the following sets of data as qualitative or quantitative.

______________ **1.** The religious affiliations of college students

______________ **2.** The height of each member of a basketball team

______________ **3.** Students' scores on the first statistics exam

______________ **4.** The color of new SUV's on a car lot

______________ **5.** The Olympic track and field world records, such as the time for the steeplechase.

Part II Classify the following sets of data as continuous or discrete.

______________ **6.** The number of students enrolled in an accounting class

______________ **7.** The number of Whirlpool refrigerators sold by Appliance Center last month

______________ **8.** The acceleration time of an automobile

______________ **9.** The temperature of a refrigerator

______________ **10.** The number of people aboard a submarine

can be ranked

Part III Identify the measurement scale as nominal, ordinal, interval, or ratio for each of the following.

interval **11.** the temperature readings in Washington, D.C.

nominal **12.** classification of computer makes

nominal **13.** college major

ordinal **14.** military rank

ratio **15.** number of vehicles produced

ratio **16.** time required to complete a crossword puzzle

ordinal **17.** order of finish in the 2003 Indianapolis 500 auto race

nominal **18.** the color of the students' hair in your statistics class

ratio 19. years in which Microsoft stock split

_____ 20. number of people at a board of directors meeting

Part IV Select the correct answer and write the appropriate letter in the space provided.

_____ 21. The collection of all possible individuals, objects, or measurements is called

a. a sample.
b. a ratio measurement.
c. an inference.
d. a population

_____ 22. Techniques used to organize, summarize, and present the data that have been collected are called

a. populations.
b. samples.
c. descriptive statistics.
d. inferential statistics.

_____ 23. An individual, measurement, or object that can appear in only one category is said to be

a. exhaustive.
b. inferential.
c. descriptive.
d. mutually exclusive.

_____ 24. Techniques used to determine something about a population, based on a sample, are called

a. descriptive statistics.
b. inferential statistics.
c. populations.
d. samples.

_____ 25. A difference between the interval scale and the ratio scale is that

a. the interval scale cannot be ranked.
b. the ratio scale does not meet the exhaustive criteria.
c. the zero point on the interval scale is arbitrary.
d. the interval scale does not meet the mutually exclusive criteria.

CHAPTER 2
DESCRIBING DATA: FREQUENCY DISTRIBUTIONS AND GRAPHIC PRESENTATION

Chapter Goals

After completing this chapter, you will be able to:

1. Organize raw data into a ***frequency distribution***.
2. Portray a frequency distribution in a ***histogram***, ***frequency polygon***, and ***cumulative frequency polygon***.
3. Present raw data using such graphical techniques as ***line charts***, ***bar charts***, and ***pie charts***.

Introduction

This chapter begins our study of ***descriptive statistics.*** Recall from Chapter 1 that when using descriptive statistics we merely describe a set of data. For example, we want to describe the entry-level salary for a select group of professions. We find that the entry-level salary for accountants is $38,000, for systems analysts $48,000, for physician's assistants $80,000, and so on. This unorganized data provides little insight into the pattern of entry-level salaries, which makes conclusions difficult.

This chapter presents a technique that is used to organize raw data into some meaningful form. It is called a ***frequency distribution.*** To better understand the main features of the data, we portray the frequency distribution in the form of a frequency polygon, a histogram, or a cumulative frequency distribution. The goal is to make tables, charts, and graphs that will quickly reveal the shape of the data.

Constructing a Frequency Distribution

A ***frequency distribution*** is a useful statistical tool for organizing a mass of data into some meaningful form.

> ***Frequency Distribution***: A grouping of data into mutually exclusive classes showing the number of observations in each.

As noted, a frequency distribution is used to summarize and organize large amounts of data.

The steps to follow in developing a frequency distribution are:

1. Decide on the number of classes.
2. Determine the class interval or width.
3. Set the individual class limits.
4. Tally the observations into the appropriate classes.
5. Count the number of items in each class.

Length of Service (in years)					
4	3	2	10	6	6
5	8	4	8	4	
6	2	3	3	7	5

As an example, the lengths of service, in years, of a sample of seventeen employees are given above.

The seventeen observations are referred to as **raw data** or **ungrouped data.** To organize the lengths of service into a frequency distribution:

1. We decide to have five classes.
2. We used a class width of 2.
3. We used classes 1 up to 3, 3 up to 5, and so on.
4. **Tally** the lengths of service into the appropriate classes.
5. Count the number of tallies in each class as shown.

Frequency Distribution

Lengths of service	Tallies	Number of employees
1 up to 3 years	//	2
3 up to 5 years	//////	6
5 up to 7 years	/////	5
7 up to 9 years	///	3
9 up to 11 yrs.	/	1
Total		17

How many classes should there be? A common guideline is from 5 to 15. Having too few or too many classes gives little insight into the data. A rule for determining the number of classes is shown on the next page. The size of the class interval may be a value such as 3, 5, 10, 15, 20, 50, 100, 1,000, and so on.

Class Interval: The size or width of the class.

The class interval can be approximated by the formula:

Class Interval $$\text{Class Interval}(i) \geq \frac{\text{highest value} - \text{lowest value}}{\text{number of classes}} \text{ or } i \geq \frac{H-L}{k}$$

Where:
i is the class interval.
H is the highest observed value.
L is the lowest observed value.
k is the number of classes.

If we apply the formula to our example, then $H = 10$, $L = 2$, and $k = 5$. We get a class interval of 2,

found by: $i \geq \frac{10-2}{5} \geq \frac{8}{5} \geq 1.6$ which is rounded to 2.

Each class has a lower class limit and an upper class limit. The lower limit of the first class is usually slightly below the smallest value and is a multiple of the class interval.

In the previous example, the smallest number of years of service is 2. We selected 1, which is slightly below 2, as the lower limit of the first class. The lower limit of the second class is 3 years, and so on.

The number of tallies or observations that occurs in each class is called the ***class frequency***.

Class frequency: The number of observations in each class.

In the example, the class frequency of the lowest class is 2. For the next higher class it is 6. The class midpoint divides a class into two equal parts.

Class midpoint: The point halfway between the upper and lower limit of a class.

The class midpoint is computed by adding the lower limit of consecutive classes and dividing the result by two. Note that the class midpoint is also called the ***class mark***.

In the example, the class midpoint of the 5 up to 7 class is 6 found by (5 + 7)/2. The class interval is the distance between the lower limit of two consecutive classes. It is 2, found by subtracting 1 (the lower limit of the first class) from 3 (the lower limit of the second class).

Suggestions on Constructing Frequency Distributions

When constructing frequency distributions, follow these guidelines:

1. *The class intervals used in the frequency distribution should be equal.* Unequal class intervals present problems in graphically portraying the distribution. However, in some situations unequal class intervals may be necessary in order to avoid a large number of empty classes.
2. The formula is based on the number of classes, and is useful for determining the class interval.

$$\text{Class Interval}\,(i) \geq \frac{\text{highest value} - \text{lowest value}}{\text{number of classes}} \text{ or } i \geq \frac{H-L}{k}$$

3. *Your professional judgment can determine the number of classes.* Too many classes or too few classes might not reveal the basic shape of the distribution. A general rule is that it is best to use at least 5 and not more than 15 classes when constructing a frequency distribution.
4. *The "2 to the k rule" is also used to determine the number of classes.* To estimate the number of classes we select the smallest integer (whole number) such that $2^k \geq n$ where n is the total number of observations. Suppose a set of data has 60 observations. If we try $k = 5$, we get $2^5 = 32$, which is less than 60, so we try $2^6 = 64$, which is greater than 60. Thus the recommended number of classes is 6. The table is based on the *"2 to the k rule."*

2 to the *k* Rule for Number of Classes	
Total Number of Observations	**Recommended Number of Classes**
9 – 16	4
17 – 32	5
33 – 64	6
65 – 128	7
129 – 256	8
257 – 512	9
513 – 1,024	10

5. *The lower limit of the first class should be an even multiple of the class interval.* Suppose a sample of weight losses ranged from 25 pounds to 64 pounds. We want to organize the weight losses into a frequency distribution with an interval of 6 pounds. The lower limit of the first class would be 24, found by multiplying 4, the even multiple, by 6, the class interval. Obviously this suggestion was not followed in the above example. Keep in mind that these are only suggestions not rules.
6. *Avoid overlapping stated class limits.* Class limits such as 4-6 and 6-8 should not be used. Use 4 up to 6, then 6 up to 8. This way you can determine in which class to tally 6.
7. *Try to avoid open-ended classes.* Open-ended classes cause serious graphing problems and make it difficult to calculate various measures described in Chapter 3.

Relative Frequency Distribution

It is often helpful to know what percent the class frequencies are of the total number of observations.

Relative class frequency: Shows what percent each class is of the total number of observations (frequencies).

The relative class frequency is found by dividing each of the class frequencies by the total number of frequencies.

Using the distribution of the lengths of service of the seventeen employees, the relative frequency for the 1 up to 3-year class is 0.1176 found by 2/17 = 0.1176 = 12%. Thus 12% of the employees had 1 up to 3 years of service. The relative frequencies for the remaining classes are shown.

Relative Frequency Distribution			
Length of service (in years)	Number of employees	Relative Frequency	Found by
1 up to 3 years	2	0.1176 = 12%	2/17
3 up to 5 years	6	0.3529 = 35%	6/17
5 up to 7 years	5	0.2941 = 29%	5/17
7 up to 9 years	3	0.1765 = 18%	3/17
9 up to 11 years	1	0.0588 = 6%	1/17
Total	17	0.9999 = 100%	

Graphic Presentation of a Frequency Distribution

To get reader attention a frequency distribution is often portrayed graphically as a histogram, a frequency polygon, and the cumulative frequency polygon.

Histogram

The simplest type of a statistical chart is called a ***histogram***.

> ***Histogram:*** A graph in which the classes are marked on the horizontal axis and the class frequencies on the vertical axis. The class frequencies are represented by the heights of the bars and the bars are drawn adjacent to each other.

For the length of service for the sample of seventeen employees a histogram would appear as shown on the right.

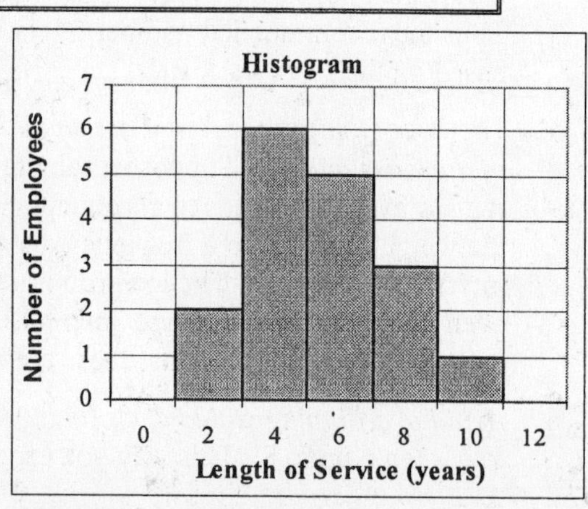

Note that to plot the bar for the 1 up to 3 years (which has a midpoint of 2 years), we drew lines vertically from 1 and from 3 years to 2 employees on the *y*-axis and then connected the end points by a straight line. The histogram provides an easily interpreted visual representation of a frequency distribution.

Frequency Polygon

A second type of chart used to portray a frequency distribution is the ***frequency polygon***.

> ***Frequency Polygon***: A graph that consists of line segments connecting the points formed by the intersection of the class midpoints and the class frequency.

For the frequency polygon, the assumption is that the observations in any class interval are represented by the class midpoint. A dot is placed at the class midpoint opposite the number of frequencies in that class. For the distribution of years of service, make the first plot by selecting 2 years on the *x*-axis (the midpoint) and then go vertically on the y-axis to 2 and place a dot. This process is continued for all classes. Then connect the dots in order.

Normal practice is to anchor the frequency polygon to the x-axis. This is accomplished by extending the lines to the midpoint of the class below the lowest class (0) and to the midpoint of the class above the highest class (12).

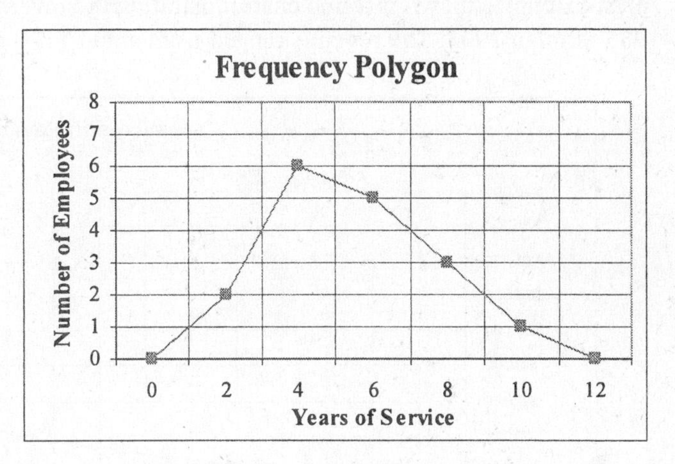

Cumulative Frequency Distributions

A ***cumulative frequency polygon*** reports the number and percent of observations that occur below a certain value.

> ***Cumulative Frequency Polygon***: A graph that shows the number of observations below a certain value.

Before we can draw a cumulative frequency polygon, we must convert the frequency distribution to a cumulative frequency distribution. To construct a cumulative frequency distribution, we add the frequencies from the lowest class to the frequency of the next highest class. We add this sum to the frequency of the next class, etc.

Cumulative Frequency Distribution

Length of service (in years)	Class Frequency	Cumulative Frequency	Found by
1 up to 3 years	2	2	2
3 up to 5 years	6	8	2 + 6
5 up to 7 years	5	13	8 + 5
7 up to 9 years	3	16	13 + 3
9 up to 11 years	1	17	16 + 1

The cumulative frequencies are plotted on the vertical axis (y-axis) and the lengths of service on the x-axis.

It may be helpful to plot the cumulative frequencies on the left side of the vertical axis and the percent of the total on the right side as shown in the polygon above.

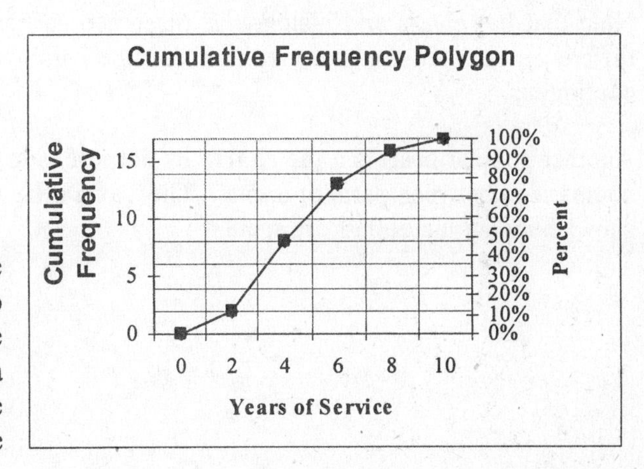

Other Graphic Presentations of Data

Several other graphic presentations of data are discussed in this section. Each is designed to emphasize certain characteristics in the data. The simple **line chart** displays information over a period of time. Time is always scaled on the horizontal axis. In the line chart a line connects the values for various periods.

As an example, shown is a line chart illustrating the revenue for the Microsoft Corporation from the years 1985 through 2004. The revenue ranged from about $140 million in 1985 to $33 billion in 2004.

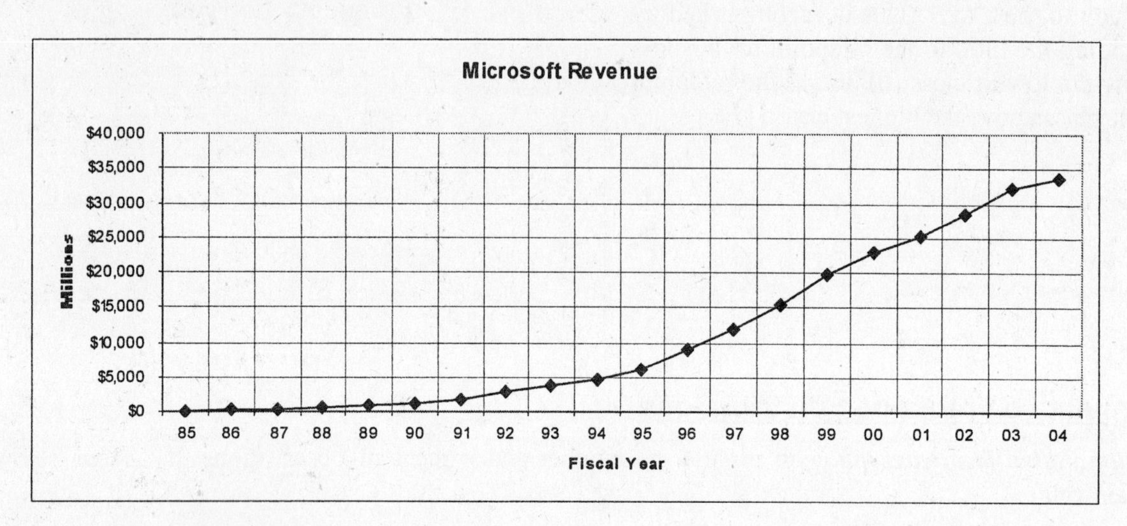

The **bar chart** is often used to display categories. For a bar chart, bars represent the data for each period. The bars can be shown as vertical or horizontal bars. As an example, shown is the prime lending rate for July, 2004 (4.25), six months ago (4.25), and a year ago (4.00).

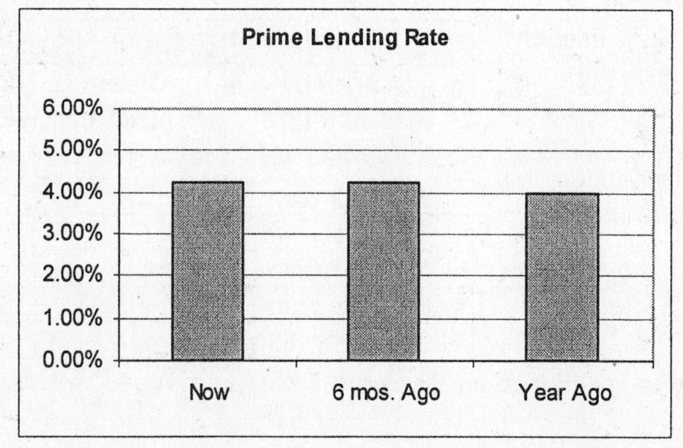

Note that bar charts and histograms, discussed earlier, both used rectangles to represent the data. The difference between the two graphs is that the bars in a histogram touch each other because the data is continuous.

Another popular chart is a **pie chart.** Its purpose is to show the relative comparison between parts of a total. The Tax Dollar Distribution table shows how our tax dollars are spent.

Tax Dollar Distribution	
Category	**Percent**
Roads	20
Education	40
Welfare	15
Salaries	18
Miscellaneous	7

After drawing a circle (pie) we put 0 on the top and go around the circle in increments of 5.

To plot the percent going for roads we draw a line from 0 to the center of the circle and another line from the center to 20. Then, 20 + 40 = 60. This area represents the amount going for education. This process is continued for the remaining items.

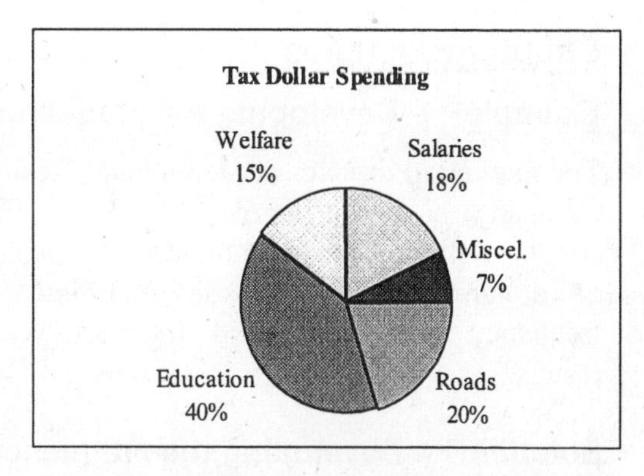

Misleading Statistics

You must be careful that you do not mislead or misrepresent your data when you construct charts and graphs. In Problems 6 and 7 we present several examples of charts and graphs that could be misleading. Whenever you see a chart or graph, study it carefully. Ask yourself: What is the writer trying to show me? Could the writer have any bias?

Glossary

Class interval: The size or width of the class.

Class frequency: The number of observations in each class.

Class midpoint: The point halfway between the upper and lower limit of a class.

Cumulative frequency polygon: A graph that shows the number of observations below a certain value.

Frequency distribution: A grouping of data into categories showing the number of observations in each mutually exclusive category.

Frequency polygon: A graph that consists of line segments connecting the points formed by the intersection of the class midpoint and the class frequency.

Histogram: A graph in which the classes are marked on the horizontal axis and the class frequencies on the vertical axis. The class frequencies are represented by the heights of the bars and the bars are drawn adjacent to each other.

Relative class frequency: Shows what percent each class is of the total number of observations (frequencies).

Note to students:

Recall that on the first page of this chapter there was a listing of the chapter goals. A brief discussion of the chapter highlights and a glossary of terms followed these goals. Now come several problems and the solution to each of the problems. They are intended to give you a detailed solution to a real-world problem, corresponding to each of the problems discussed.

Chapter Examples

Example 1 – Developing the Frequency Distribution

The marketing director at Gomminger Realty Company selected a sample of 30 homes for study. (Selling price is reported in thousands of dollars.) Organize these data into a frequency distribution and interpret your results.

$76	$74	$71	$78	$80	**$67**	←Low
80	82	67	88	72	78	
85	76	84	82	83	80	
72	82	77	79	86	**89**	←High
69	70	82	86	78	77	

Solution 1 – Developing the Frequency Distribution

Step 1: Decide the number of classes.

Determine the number of classes by using the "*2 to the k rule*". To estimate the number of classes, select the smallest integer (whole number) such that $2^k \geq n$ where n is the total number of observations. Our set of data has 30 observations. If we try $k = 5$, *the answer is* $2^5 = 32$, which is more than 30. Thus the recommended number of classes is 5.

Selling Price ($000)	Tallies	Number of Homes
$65 up to $70	///	3
$70 up to $75	~~////~~	5
$75 up to $80	~~////~~ ///	8
$80 up to $85	~~////~~ ////	9
$85 up to $90	~~////~~	5
		30

Step 2: Determine the class interval or width.

Observe that the home with the lowest selling price was $67 thousand and the highest was $89 thousand. Use the formula to determine the interval.

$$\text{Class Interval}(i) = \frac{\text{highest value} - \text{lowest value}}{\text{number of classes}} \text{ or } i \geq = \frac{H-L}{k} = \frac{89\text{-}67}{5} = \frac{22}{5} = 4.4$$

We round 4.4 up to 5, thus we let the class interval be $5 thousand.

Step 3: Set the individual class limit.

We decide to let $65 thousand be the lower limit of the first class. Thus, the first class will be $65 up to $70 thousand and the second class $70 up to $75 thousand, and so on.

Step 4: Tally the selling prices into each of the classes.

The first home sold for $76 thousand, therefore, the price is tallied into the $75 thousand up to $80 thousand class. The procedure is continued, resulting in the frequency distribution shown above. Observe that the largest concentration of the data is in the $80 up to $85 thousand class. The procedure is continued resulting in the frequency distribution shown above.

Step 5: Count the number of items in each class.

As noted before, the class frequencies are the number of observations in each class. For the $65 up to $70 thousand class the class frequency is 3, and for the $70 up to $75 thousand class the class frequency is 5. This indicates that three homes sold in the $65 up to $70 thousand price range and five in the $70 up to $75 thousand range.

It is also clear that the interval between the lowest and highest selling price in each category is $5 thousand. How would we classify a home selling for $70 thousand? It would fall in the second class. Homes selling for $65,000 up through $69,999.99 go in the first class, but a home selling for more than this amount goes in the next class. So the $70,000 selling price puts the home in the second class.

The class midpoint is determined by going halfway between the lower limit of consecutive classes. Halfway between $65 and $70 is $67.5 thousand, the class midpoint.

Self-Review 2.1

Check your answers against those in the ANSWER section.

This is the first in a series of exercises designed to check your comprehension of the material just presented. We suggest that you work all parts of the exercise. Then check your answers against those given in the answer section of this study guide.

The Jansen Motor Company has developed a new engine to further reduce gasoline consumption. The new engine was installed in 20 mid-sized cars and the number of miles per gallon recorded (to the nearest mile per gallon).

29	32	20	30	39
27	28	21	36	20
27	18	32	37	29
30	23	25	19	30

a. Use the "2 to the k rule" to determine the number of classes.
b. Determine the class interval.
c. Develop a frequency distribution.

Example 2 – Developing the Frequency Distribution

Based on the information from Gomminger Realty in Example 1,

a. Develop a relative frequency distribution.

b. What percent of the homes sold for a price from $85,000 up to $90,000?

Solution 2 – Developing the Frequency Distribution

a. Relative class frequencies show the fraction of the total number of observations in each class. These fractions are often expressed as percents. To convert a frequency distribution to a relative frequency distribution, each class frequency is divided by the total number of observations.

Using the distribution from Example 1, the relative frequency for the $65 up to $70 class is 0.10, found by dividing 3 by 10. Thus, 10% of the homes sold from $65,000 up to $75,000.

All the relative frequencies are shown in the table.

Selling Price ($000)	Tallies	Number of Homes	Relative Frequency	Found by
$65 up to $70	///	3	0.1000 = 10.00%	3/30
$70 up to $75	~~////~~	5	0.1667 = 16.67%	5/30
$75 up to $80	~~////~~ ///	8	0.2667 = 26.67%	8/30
$80 up to $85	~~////~~ ////	9	0.3000 = 30.00%	9/30
$85 up to $90	~~////~~	5	0.1667 = 16.67%	5/30
Total		30	1.0001 = 100.01%*	
*Note that rounding caused the relative frequency to exceed 100%				

b. The relative frequency is 16.67%, which means 16.67% of the homes sold for a price from $85,000 up to $90,000.

Self-Review 2.2

Check your answers against those in the ANSWER section.

Use the Jansen Motor Company data in Self-Review 2.1 to construct a relative frequency distribution.

Example 3 – Developing a Histogram

Based on the information from Gomminger Realty in Example 1, develop a histogram.

Solution 3 – Developing a Histogram

The class frequencies are scaled on the vertical axis (y-axis) and the selling price on the horizontal (x-axis). A vertical line is drawn from the two class limits of a class to a height corresponding to the number of frequencies. The tops of the lines are then connected.

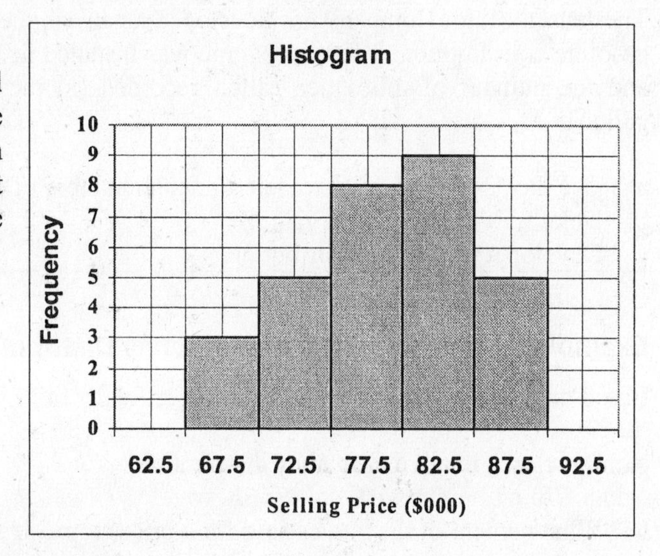

Self-Review 2.3

Check your answers against those in the ANSWER section.

Use the Jansen Motor Company data in Self-Review 2.1 to construct a histogram.

Example 4 – Construct a Frequency Polygon

Use the information contained in Examples 1 and 2 to construct a frequency polygon.

Solution 4 – Construct a Frequency Polygon

Class frequencies are scaled on the vertical axis (y-axis) and class midpoints along the horizontal axis (x-axis).

1. The first plot is at point 67.5 on the x-axis and 3 on the y-axis.
2. The midpoints of the class below the first class and above the last class are added. This allows the graph to be anchored to the x-axis at zero frequencies.

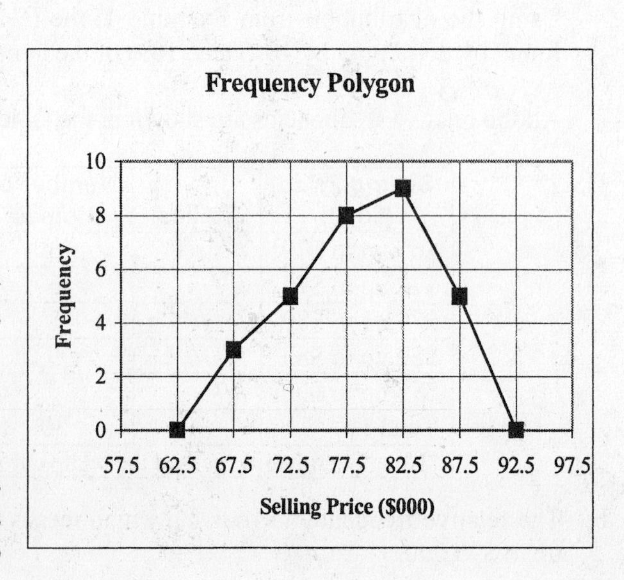

Self-Review 2.4

Check your answers against those in the ANSWER section.

Use the Jansen Motor Company data in Self-Review 2.1 to construct a frequency polygon.

Example 5 – Develop a Cumulative Frequency Polygon

Based on the information in Example 1

a. Construct a cumulative frequency polygon.
b. Estimate the price below which 75 percent of the homes were sold.
c. Estimate the number of homes sold for less than $72,000.

Class limits ($000)	Class Frequency	Cumulative Frequency
$65 up to $70	3	3
$70 up to $75	5	8
$75 up to $80	8	16
$80 up to $85	9	25
$85 up to $90	5	30

Solution 5 – Develop a Cumulative Frequency Polygon

Construct a cumulative frequency distribution by using the class limits. The first step is to determine the number of observations "less than" the upper limit of each class. Three homes were sold for less than $70 and eight were sold for between $65 and $75 thousand. The eight is found by adding the three that sold for $65 to $70 thousand and the five that sold for between $70 and $75 thousand. The cumulative frequency for the fourth class is obtained by adding the frequencies of the first four classes. The total is 25, found by (3 + 5 + 8 + 9). The less-than-cumulative frequency distribution would appear as shown.

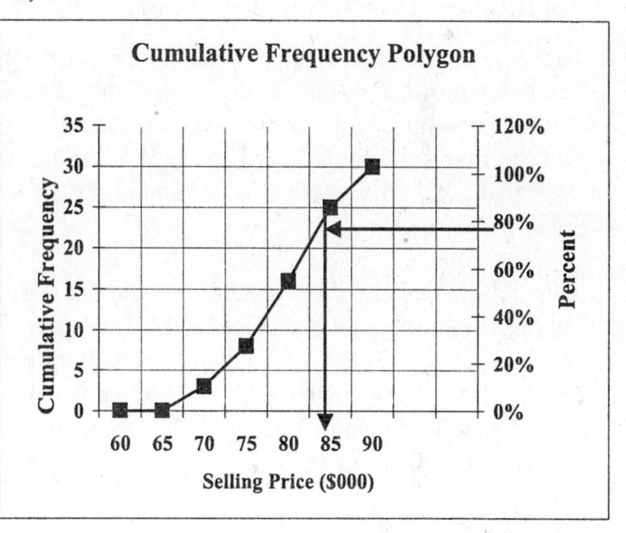

a. To construct a cumulative frequency polygon the upper limits are scaled on the x-axis and the cumulative frequencies on the y-axis. The cumulative percents are placed along the right-hand scale (vertical). The first plot is $x = 70$ and $y = 3$. The next plot is 75 and 8. As shown, the points are connected with straight lines (see the above chart).

b. To estimate the amount for which less than 75 percent of the homes were sold, draw a horizontal line from the cumulative percent (75) over to the cumulative frequency polygon. At the intersection, draw a line down to the x-axis giving the approximate selling price. It is about $85 thousand. Thus, about 75 percent of the homes sold for $85,000 or less.

c. To estimate the percent of the homes that sold for less than $74,000, first locate the value of $74 on the x-axis. Next, draw a vertical line from the x-axis at 74 up to the graph. Draw a line horizontally to the cumulative percent axis and read the cumulative percent. It is about 18%. Hence, we conclude that about 18 percent of the homes were sold for less than $74,000.

> **Self-Review 2.5**
>
> Check your answers against those in the ANSWER section.
>
> Use the Jansen Motor Company data in Self-Review 2.1.
>
> a. Construct a cumulative frequency polygon.
> b. Estimate the percent of the automobiles getting less than 30 miles per gallon.
> c. Twenty percent of the automobiles obtain how many miles per gallon or less?

Example 6 – Drawing a Line Chart

The percent of disposable income (disposable income is the amount of income left after taxes) spent for groceries for the period from 1975 to 2005 is shown on the right. Draw a line chart to depict the trend.

Year	Disposable Income spent on groceries
1975	13.0%
1980	12.3%
1985	11.2%
1990	10.1%
1995	9.5%
2000	9.3%
2005	9.0%

Solution 6 – Drawing a Line Chart

The time is scaled at five-year intervals on the horizontal or *x*-axis. The percent of disposable income spent for groceries is scaled on the vertical or *y*-axis. Two different versions are shown. In each version plot the first point by going up from 1975 on the *x*-axis to 13%. Plot the second point by going up from 1980 to 12.3%. This process is continued for the remaining periods. The dots are connected with straight lines.

Note that in Version 2 the vertical axis did not start from zero. Technically this is called a scale break. That is, we started at 8 and ended at 14. In Version 1 we scaled the vertical axis from 0 to 14. Both versions are correct and indicate the trend for spending disposable income for groceries; however, the visual impact is somewhat different. Notice the change in emphasis. Version 2 "shows" a more dramatic decline than is shown in Version 1. The more dramatic decline is brought about because of the use of the scale break.

Version 1

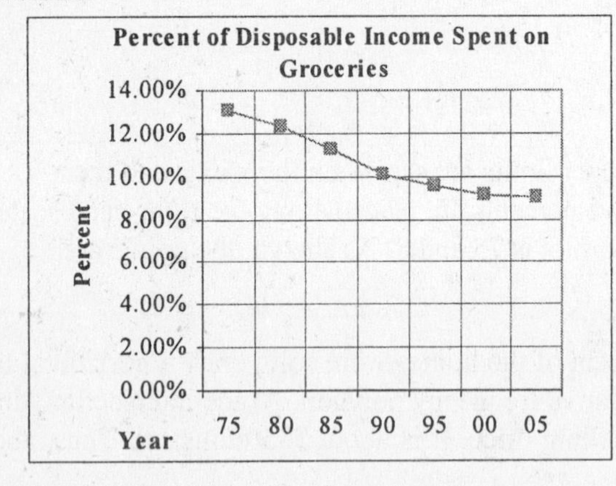

Version 2 – Misleading

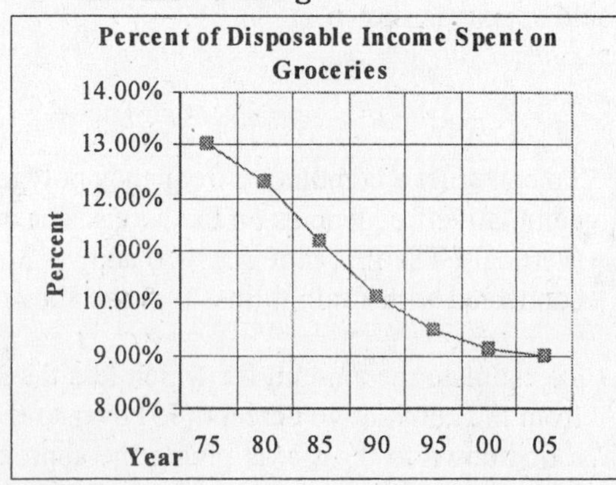

Self-Review 2.6

Check your answers against those in the ANSWER section.

The expenditures on research and development for the Hennen Manufacturing Company are given. Construct a simple line chart.

Year	Expenditure ($000)
1999	94
2000	103
2001	115
2002	145
2003	175
2004	203
2005	190

Example 7 – Drawing a Simple Bar Chart

Refer to Example 6. Develop a simple bar chart for the percent of disposable income spent for groceries.

Solution 7 – Drawing a Simple Bar Chart

The usual practice is to scale time along the horizontal axis. The height of the bars corresponds to percent of disposable income spent for groceries. Two different versions are shown. In each version, to form the first bar, draw parallel vertical lines from 1975 up to 13.0%. Draw a line parallel to the *x*-axis at 13.0% to connect the lines. This process is continued for the other periods. Note that in Version 2 the vertical axis did not start from zero. Technically this is called a scale break. That is, we started at 8 and ended at 14. In Version 1 we scaled the vertical axis from 0 to 14. Both versions are correct and indicate the trend for spending disposable income for groceries. However, the visual impact is somewhat different. Notice the change in emphasis. As noted on the previous page, Version 2 "shows" a more dramatic decline than is shown in Version 1. The more dramatic decline is brought about because of the use of the scale break.

Version 1

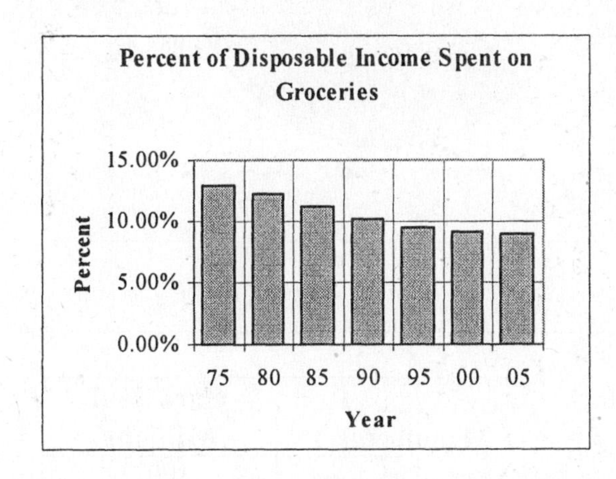

Version 2 – Misleading

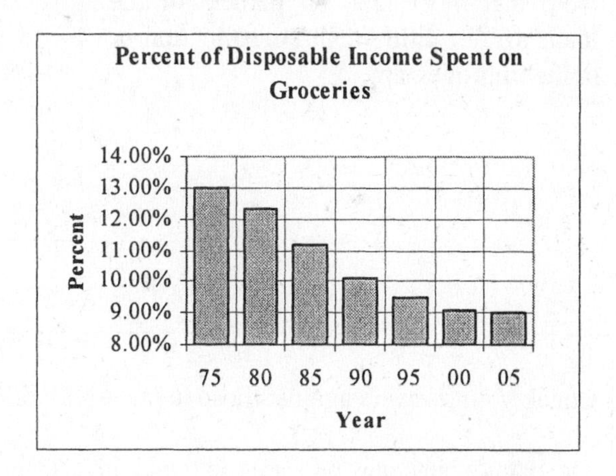

Self-Review 2.7

Check your answers against those in the ANSWER section.

Use the Hennen Manufacturing data of Self-Review 2.6 to construct a simple bar chart.

Example 8 – Drawing a Pie Chart

The table to the right reports the purpose of home equity loans by the Home Bank and the percent each type of loan is relative to the total. Portray the home equity loan information in the form of a pie chart.

Loan Purpose	Percent Of Total	Cumulative Percent
Home improvement	32	32
Debt consolidation	30	62
Car purchase	11	73
Education	10	83
Other	9	92
Investments	8	100

Solution 8 – Drawing a Pie Chart

The first step is to draw a circle. Next draw a line from 0 to the center of the circle and another from the center of the circle to 32%. Adding the 32% for home improvements and the 30% for debt consolidation gives 62%. A line is drawn from the center to 62%. The area between 32% and 62% represents the percent of equity loans for debt consolidation. The process is continued for the remaining cumulative percents. Note that more than 60 percent of the loans are for either debt consolidation or home improvement.

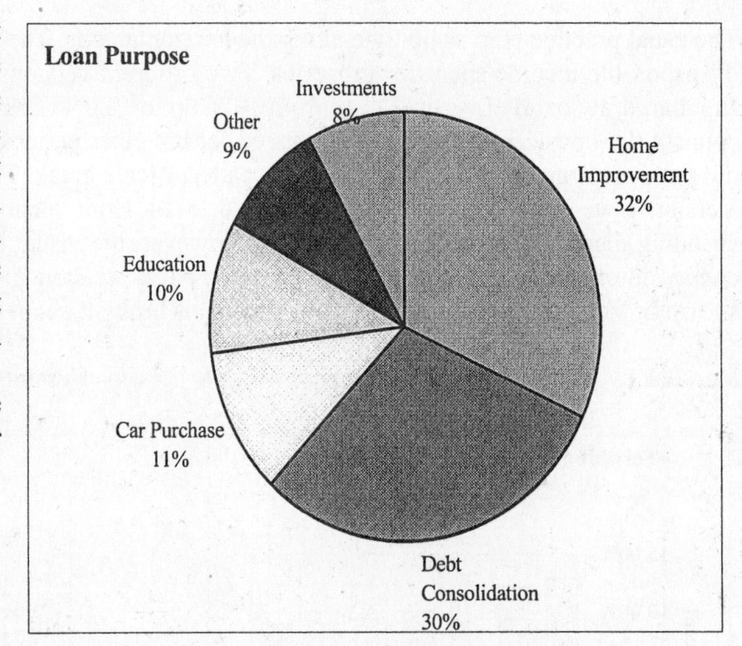

Self-Review 2.8

Check your answers against those in the ANSWER section.

The data depicts new cars sold in the United States for the year, classified by manufacturer. Portray the "new cars sold" data in the form of a pie chart.

Manufacturer	Cars Sold Millions
General Motors	3100
Ford	1900
DaimlerChrysler	800
Toyota	800
Honda	800
Nissan	500
Other	1100
Total	9000

CHAPTER 2 ASSIGNMENT

DESCRIBING DATA: FREQUENCY DISTRIBUTIONS AND GRAPHIC PRESENTATION

Name ______________________________ Section __________ Score __________

Part I Select the correct answer and write the appropriate letter in the space provided

__b__ **1.** A grouping of data into classes giving the number of observations in each class is called a(an)
- **a** bar chart.
- **b.** frequency distribution.
- **c.** pie chart.
- **d.** cumulative frequency distribution.

__d__ **2.** The distance between consecutive lower class limits is called the
- **a.** class interval.
- **b.** frequency distribution.
- **c.** class midpoint.
- **d.** class frequency.

__b__ **3.** The class midpoint is
- **a.** equal to the number of observations.
- **b.** found by adding the upper and lower class limit and dividing by 2.
- **c.** equal to the class interval.
- **d.** all of the above.

__c__ **4.** The number of observations in a particular class is called the
- **a.** class interval.
- **b.** class frequency.
- **c.** frequency distribution.
- **d.** none of the above.

__b__ **5.** A bar chart is used most often when
- **a.** you want to show frequencies as compared to total observations.
- **b.** you want to show frequencies by class intervals.
- **c.** you want to display frequencies by category.
- **d.** you want to organize data along certain time interval.

__c__ **6.** In a *relative frequency* distribution
- **a.** the class frequencies are divided by 100.
- **b.** the data are related to each other rather than mutually exclusive.
- **c.** the class frequency is divided by the total number of observations.
- **d.** the frequencies are added together to give a relative set of numbers.

__c__ **7.** For a line chart involving time in years and dollar values, the horizontal or *X*-axis would be used to represent
- **a.** the dollar variable.
- **b.** the time variable.
- **c.** the class interval.
- **d.** the class frequency.

___b___ 8. The suggested interval size of the class intervals for a histogram can be estimated by:

a. consecutive lower class limits divided by 2.
b. consecutive lower class limits divided by the total number of observations.
c. using the formulas: $i \geq \frac{H - L}{k}$
d. consecutive lower class limits divided by the number of frequencies in each class.

___a___ 9. A pie chart requires at least what level of data?

a. nominal
b. ordinal
c. interval
d. ratio

___d___ 10. A graphic representation of a frequency distribution constructed by connecting the class midpoints with lines is called a

a. histogram.
b. line chart.
c. pie chart.
d. frequency polygon.

Part II Show all of your work. Write the answer in the space provided.

11. Shown below are the net sales for the J. M. Smucker Company, a leading marketer of jams and jellies. Use the data to construct a line graph.

Smucker's Net Sales	
Year	**Sales (millions)**
1990	345
1991	399
1992	425
1993	454
1994	462
1995	478
1996	511
1997	529
1998	524
1999	565
2000	602
2001	632
2002	651
2003	687
2004	1,311

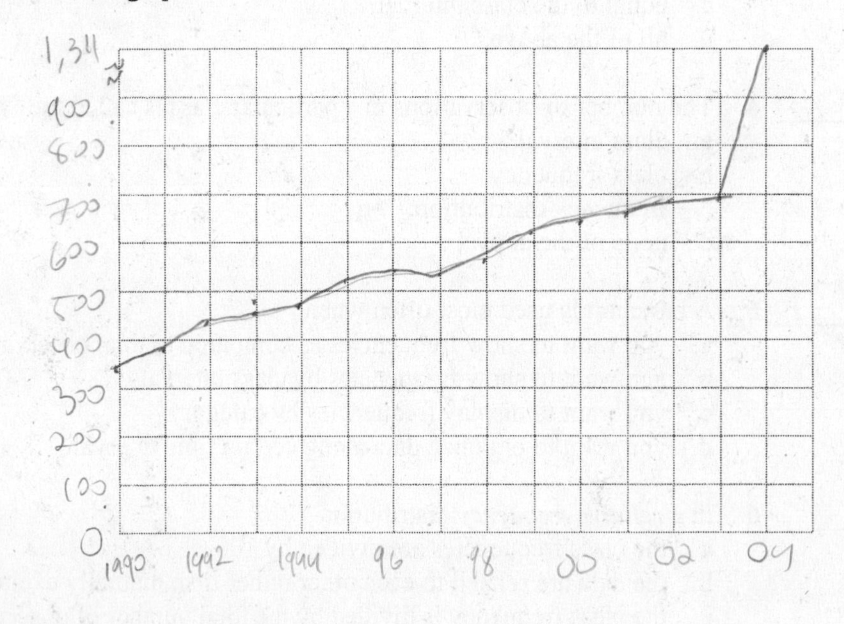

12. The following is a breakdown of the expenditures of the Ohio Division of Wildlife for 2004. Construct a pie chart.

Category	Amount (millions)
Administration	2.0
Education	4.7
Law enforcement	3.6
Wildlife officers	7.1
Fish management	7.6
Wildlife management	10.5
Operations	7.6
Capital improvements	2.0

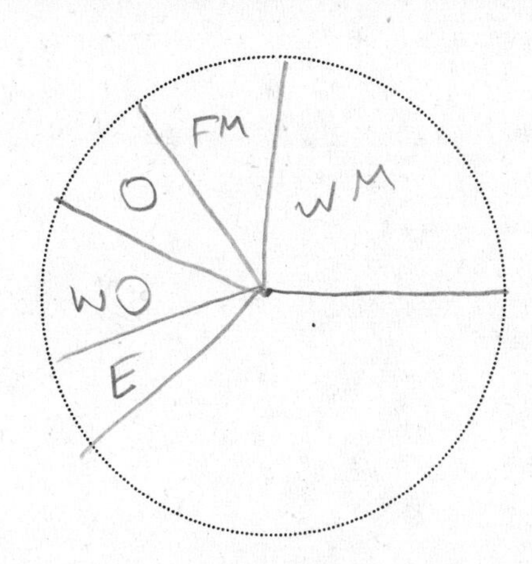

13. Listed are the weights of the 2004 Super Bowl Champion New England Patriots starting lineup, including the place kicker and the punter. Organize the data into

228	209	195	305	324	215
241	291	181	242	234	320
190	210	230	263	194	205
326	333	186	225	279	255

a. a frequency distribution (use a class interval of 30, with 180 as the lower limit of the first class)
b. a relative frequency distribution
c. a cumulative frequency distribution.

d. Draw a histogram for the data.

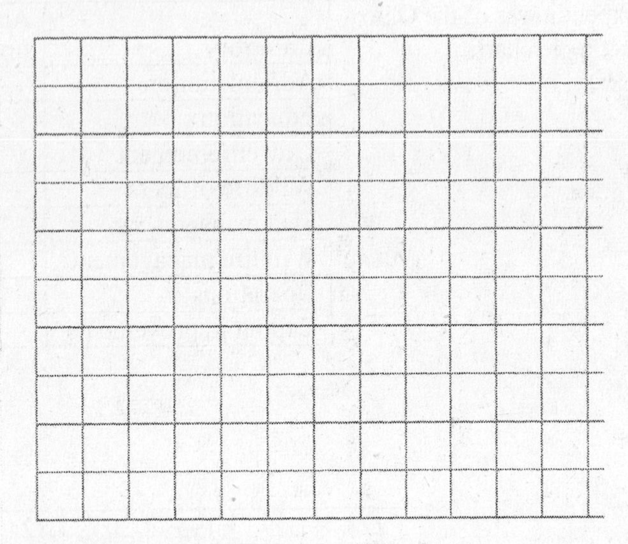

e. Develop a frequency polygon.

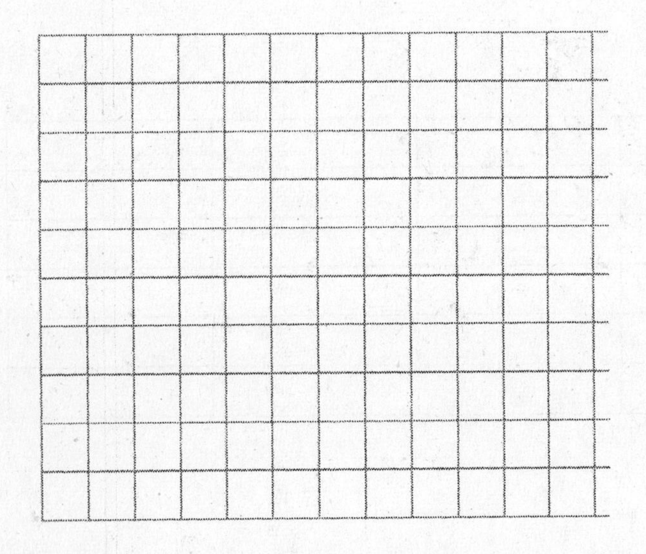

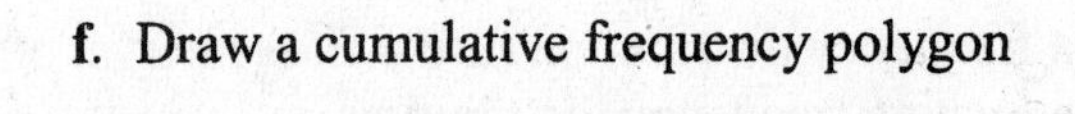

f. Draw a cumulative frequency polygon

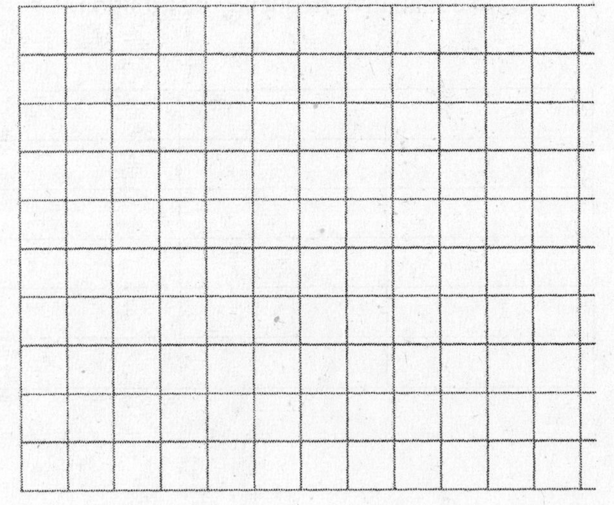

CHAPTER 3
DESCRIBING DATA: NUMERICAL MEASURES

Chapter Goals

After completing this chapter, you will be able to:

1. Calculate the ***arithmetic mean***, ***weighted mean***, ***median***, ***mode***, and ***geometric mean***.
2. Explain the characteristics, uses, advantages, and disadvantages of each ***measure of location***.
3. Identify the position of the mean, median, and mode for both ***symmetric*** and ***skewed distributions***.
4. Compute and interpret the ***range***, ***mean deviation***, and ***standard deviation***.
5. Understand the characteristics, uses, advantages, and disadvantages of each ***measure of dispersion***.
6. Understand ***Chebyshev's*** theorem and ***Empirical Rule*** as they relate to a set of observations.

Introduction

What is an average? It is a single number used to describe the central tendency of a set of data.

Examples of an average are:

- The average length of the school year for students in public schools in the United States is 180 days.
- The median salary of New York Yankees baseball players on opening day 2004 was $3,100,000. (***http://asp.usatoday.com/sports/baseball/salaries/mediansalaries.aspx?year=2004***)
- The median price of houses on Kelley's Island went up 51% in five years to $208,250. (***The Blade***, June 27, 2004)
- "The 2004 average income for Computer Support Specialists was $20.50 per hour. (***The Census Bureau's Income Statistics Branch***, July, 2004)
- Computer Software Engineers' pay averaged $36.65 per hour, while Sales Cashiers averaged $7.58 per hour. (***Bureau of Labor Statistics web site: http://stats.bls.gov, July, 2004***)

There are several types of averages. We will consider five: the arithmetic mean, the median, the mode, the weighted mean, and the geometric mean.

Measures of Location

The purpose of a measure of location is to pinpoint the center of a set of observations.

> ***Measure of location***: A single value that summarizes a set of data. It locates the center of the values.

The arithmetic mean, or simply the mean, is the most widely used measure of location.

> ***Mean***: The sum of observations divided by the total number of observations.

The population mean is calculated as follows:

$$\text{Population mean} = \frac{\text{Sum of all values in the population}}{\text{Number of values in the population}}$$

In terms of symbols, the formula for the mean of a population is:

Population Mean $$\mu = \frac{\Sigma X}{N} \qquad [3-1]$$

Where:

μ represents the population mean. It is the Greek letter "mu."
N is the number of items in the population.
X is any particular value.
$\sum$ indicates the operation of adding all the values. It is the Greek letter "sigma."
$\sum X$ is the sum of the X values.
[3-1] indicates the formula number from the text.

Any measurable characteristic of a population is called a ***parameter***.

Parameter: A characteristic of a population.

The Sample Mean

As explained in Chapter 1, we frequently select a sample from the population to find out something about a specific characteristic of the population.

The mean of a sample and the mean of a population are computed in the same way, but the shorthand notation is different.

In terms of symbols, the formula for the mean of a sample is:

Sample Mean $$\overline{X} = \frac{\Sigma X}{n} \qquad [3-2]$$

Where:

$\overline{X}$ is the sample mean; it is read "*X* bar".
n is the number of values in the sample.
X is a particular value.
$\sum$ indicates the operation of adding all the values.
$\sum X$ is the sum of the X values.
[3-2] is the formula number from the text.

The mean of a sample, or any other measure based on sample data, is called a ***statistic***.

Statistic: A characteristic of a sample.

"The mean weight of a sample of laptop computers is 6.5 pounds," is an example of a statistic.

In formulas [3-1] and [3-2] the mean is calculated by summing the observations and dividing by the total number of observations.

Suppose the Kellogg Company's quarterly earnings per share for the last five quarters are: $0.89, $0.77, $1.05, $0.79, and $0.95. If the earnings are a population, the mean is found by:

$$\mu = \frac{\Sigma X}{N} = \frac{(\$0.89 + \$0.77 + \$1.05 + \$0.79 + \$0.95)}{5}$$
$$= \frac{\$4.45}{5} = \$0.89$$

The mean quarterly earning per share is $0.89.

In some situations the mean may not be representative of the data.

As an example, the annual salaries of five vice presidents at AVX, LLC are $90,000, $92,000, $94,000, $98,000, and $350,000. The mean is:

$$\mu = \frac{\Sigma X}{N} = \frac{(\$90,000 + \$92,000 + \$94,000 + \$98,000 + \$350,000)}{5}$$
$$= \frac{\$724,000}{5} = \$144,800$$

Notice how the one extreme value ($350,000) pulled the mean upward. Four of the five vice presidents earned less than the mean, raising the question whether the arithmetic mean value of $144,800 is typical of the salary of the five vice presidents.

Properties of the Mean

As stated, the mean is a widely used measure of location. It has several important properties.

1. Every set of interval level and ratio level data has a mean.
2. All the data values are included in the calculation.
3. A set of data has only one mean, that is, the mean is unique.
4. The mean is a useful measure for comparing two or more populations.
5. The sum of the deviations of each value from the mean will always be zero, that is:

$$\Sigma(X - \overline{X}) = 0$$

Weighted Mean

The ***weighted mean*** is a special case of the arithmetic mean. It is often useful when there are several observations of the same value.

> ***Weighted mean***: The value of each observation is multiplied by the number of times it occurs. The sum of these products is divided by the total number of observations to determine the weighted mean.

In general, the weighted mean of a set of values, designated $X_1, X_2, X_3, \ldots X_n$, with the corresponding weights $w_1, w_2, w_3, \ldots, w_n$ is computed by:

Weighted Mean

$$\overline{X}_w = \frac{w_1X_1 + w_2X_2 + w_3X_3 + \cdots + w_nX_n}{w_1 + w_2 + w_3 + \cdots + w_n} \qquad [3-3]$$

The weighted mean is particularly useful when various classes or groups contribute differently to the total. For example, a small accounting firm consists of administrative assistants who are paid $12 per hour, financial assistants who earn $15 per hour, and tax examiners who earn $24 per hour.

To say the average hourly wage for the firm is $16 per hour ($12 + $15 + $21) ÷ 3 would not be accurate unless there was the same number of people in each group.

Suppose the accounting firm has ten employees: two administrative assistants who earn $12 per hour, 3 financial assistants who earn $15 per hour, and five tax examiners who earn $24 per hour. The weighted mean is:

$$\overline{X}_w = \frac{w_1X_1 + w_2X_2 + w_3X_3 + \cdots + w_nX_n}{w_1 + w_2 + w_3 + \cdots + w_n}$$

$$= \frac{(2\times\$12) + (3\times\$15) + (5\times\$24)}{2+3+5} = \frac{\$24 + \$45 + \$120}{10} = \frac{\$189}{10} = \$18.90$$

Thus the weighted mean is $18.90.

The Median

It was pointed out that the arithmetic mean is often not representative of data with extreme values. The ***median*** is a useful measure when we encounter data with an extreme value.

> ***Median***: The midpoint of the values after all observations have been ordered from the smallest to the largest, or from largest to smallest.

Fifty percent of the observations are above the median and 50 percent are below the median. To determine the median, the values are ordered from low to high, or high to low, and the middle value selected. Hence, half the observations are above the median and half are below it. For the executive incomes, the middle value is $44,000, the median.

$40,000 $42,000 $44,000 $48,000 $300,000

⇑

median

Obviously, it is a more representative value in this problem than the mean of $94,800.

Note that there were an odd number of executive incomes (5). For an odd number of ungrouped values we just order them and select the middle value. To determine the median of an even number of ungrouped values, the first step is to arrange them from low to high as usual, and then determine the value half way between the two middle values.

As an example, the number of bronze castings produced in a day at Markey Bronze is 87, 62, 91, 58, 99, and 85. Ordering these from low to high:

58 62 85 87 91 99
⇑ ⇑

The median number produced is halfway between the two middle values of 85 and 87. The median is 86. Thus we note that the median (86) may not be one of the values in a set of data.

Properties of the Median

The major properties of the median are:

1. The median is a unique value, that is, like the mean, there is only one median for a set of data.
2. It is not influenced by extremely large or small values and is therefore a valuable measure of location when such values do occur.
3. It can be computed for ratio level, interval level, and ordinal-level data.
4. Fifty percent of the observations are greater than the median and fifty percent of the observations are less than the median.

The Mode

A third measure of location is the ***mode***.

> ***Mode***: The value of the observation that appears most frequently.

The mode is the value that occurs most often in a set of raw data. The dividends per share declared on five stocks were: $3, $2, $4, $5, and $4. Since $4 occurred twice (the most frequent value), the mode is $4.

Properties of the Mode

1. The mode can be found for all levels of data (nominal, ordinal, interval, and ratio).
2. The mode is not affected by extremely high or low values.
3. A set of data can have more than one mode. If it has two modes, it is said to be bimodal.
4. A disadvantage is that a set of data may not have a mode because no value appears more than once.

The Relative Positions of the Mean, Median, and Mode

The mean, median, and mode of a set of data are usually not all equal. However, if they are identical, the distribution is a ***symmetrical distribution***.

> ***Symmetrical distribution***: A distribution that has the same shape on either side of the median.

The chart on the right shows the useful life of a sample of batteries used in a CD player. Note the symmetrical bell-shape of the distribution. In a symmetrical distribution the mean, median and mode are equal.

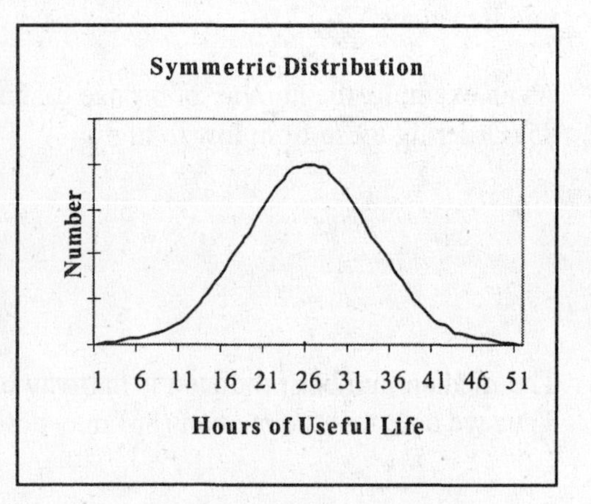

If the distribution is not symmetrical, it is skewed and the relationship between the mean, median, and mode changes. If the long tail is to the right, the distribution is said to be a ***positively skewed distribution***.

> ***Positively skewed distribution***: The long tail is to the right; that is, in the positive direction. The mean is larger than the median or the mode.

The chart on the right shows the years of service for a group of employees at an old manufacturing plant that was revitalized with a new product line and experienced a hiring surge about 13 years ago. It is a positively skewed distribution. The mean is larger than the median, which is larger than the mode.

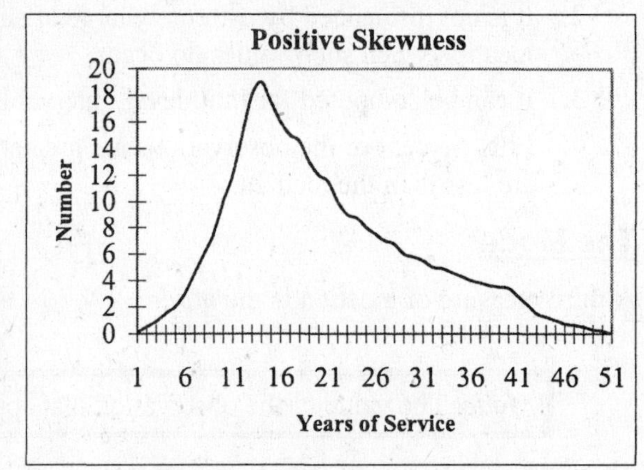

For a ***negatively skewed distribution*** the mean is the smallest of the three measures of central tendency (because it is being pulled down by the small observations). The mode is the highest of the three measures.

> ***Negatively skewed distribution***: The long tail is to the left or in the negative direction. The mean is smaller than the median or mode.

The chart on the right shows the years of service for a group of teachers in a school system that has an experienced staff and has not hired many staff in recent years. The mean is smaller than the median, which is smaller than the mode.

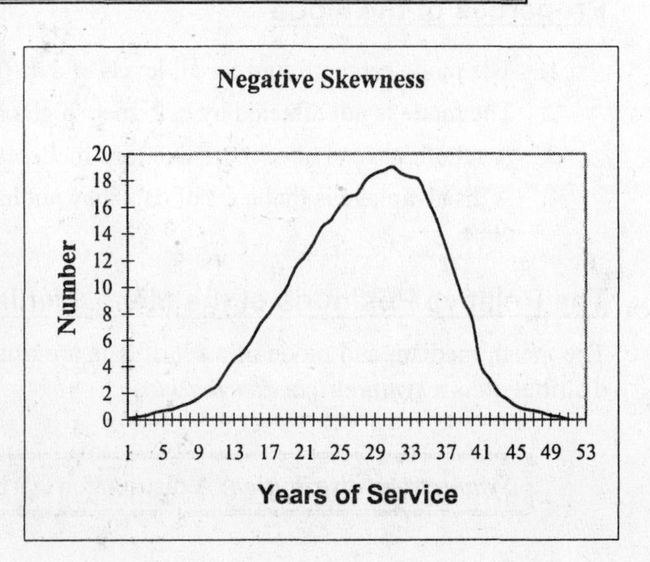

In skewed distributions the mode always appears at the apex or top (highest point) on the curve and the mean is pulled in the direction of the tail. The median always appears between the mode and the mean, regardless of the direction of the tail.

The Geometric Mean

The geometric mean is used to determine the mean percent increase from one period to another. It is also used in finding the average of ratios, indexes, and growth rates.

Geometric mean: The nth root of the product of n values.

The formula for finding the geometric mean is:

Geometric Mean $$GM = \sqrt[n]{(X_1)(X_2)(X_3)\cdots(X_n)} \qquad [3-4]$$

Where:

X_1, X_2, (X_3) etc.	are data values.
n	is the number of values.
$\sqrt[n]{\ }$	is the n th root.

The geometric mean can be used for averaging percents. Suppose the return on investment for McDermoll International for the past 4 years is 0.4%, 2.9%, 2.1%, and 12.3%. The GM increase over the period is 4.3 percent, found by: [Note that the 1.004 represents the 0.4% return on investment plus the original investment of 1.000. This is also done for the other returns]

$$GM = \sqrt[n]{(X_1)(X_2)(X_3)\cdots(X_n)}$$
$$= \sqrt[4]{1.004 \times 1.029 \times 1.021 \times 1.123}$$
$$= \sqrt[4]{1.18455} = 1.043$$

The geometric mean is fourth root of 1.18455, which is 1.043. The average return on the investment is found by subtracting one from the geometric mean. (1.043 – 1.000) = 0.043 = 4.3%.

Another application of the geometric mean is to find average percent increase over a period of time. Text formula [3-5] is used:

Average Percent Increase Over Time $$GM = \sqrt[n]{\frac{\text{Value at end of period}}{\text{Value at beginning of period}}} - 1 \qquad [3-5]$$

Why Study Dispersion?

A direct comparison of two sets of data based only on two measures of location such as the mean and the median can be misleading since an average does not tell us anything about the spread of the data.

For example, the mean salary paid to baseball players for the New York Yankees is $6,568,757. However, the range is $21,697,450, with a low of $302,550 and a high of $22,000,000. The Tampa Devil Rays have a mean salary of $1,094,681. The range is $7,197,500, with a low of $302,500 and a high of $7,500,000. (***http://espn.go.com/mlb/clubhouses/salaries/2004***).

Suppose a statistics instructor has two classes, one in the morning and one in the evening; each with six students. In the morning class (AM) the students' ages are 18, 20, 21, 21, 23, and 23 years. In the evening class

(PM) the ages are 17, 17, 18, 20, 25, and 29 years. Note that for both classes the mean age is 21 years but there is more variation or dispersion in the ages of the evening students.

A small value for a measure of dispersion indicates that the data are clustered closely, say, around the arithmetic mean. Thus the mean is considered representative of the data, that is, it is reliable. Conversely, a large measure of dispersion indicates that the mean is not reliable and is not representative of the data.

Measures of Dispersion

We will consider several measures of dispersion: the ***range***, the ***mean deviation***, the ***variance***, and the ***standard deviation***.

Range

The simplest measure of dispersion is the ***range***.

> ***Range***: The difference between the largest and smallest values in a data set.

The formula for range is:

Range	Range = Largest value – Smallest value	[3-6]

The statistics instructor referred to above has two classes with the ages indicated:

A.M. Class: 18, 20, 21, 21, 23, 23 **P.M. Class**: 17, 17, 18, 20, 25, 29

The range for the classes is:

A.M. Class: (23 – 18) = 5 **P.M. Class**: (29 – 17) = 12

Thus we can say that there is more spread in the ages of the students enrolled in the evening (P.M.) class compared with the morning (A.M.) class.

The characteristics of the range are:

- Only two values are used in the calculation.
- It is influenced by extreme values.
- It is easy to compute and understand.
- It can be distorted by an extreme value.

The range has two disadvantages. It can be distorted by a single extreme value. Suppose the same statistics instructor has a third class of five students. The ages of these students are given in the table.

Ages of Students				
20	20	21	22	60

The range of ages is 40 years, yet four of the five students' ages are within two years of each other. The 60-year old student has distorted the spread. Another disadvantage is that only two values, the largest and the smallest, are used in its calculation.

Mean Deviation

In contrast to the range, the ***mean deviation*** considers all the data.

> ***Mean Deviation***: The arithmetic mean of the absolute values of the deviations from the arithmetic mean.

In terms of symbols, the formula for the mean deviation is:

Mean Deviation $$MD = \frac{\Sigma|X - \bar{X}|}{n} \quad [3-7]$$

Where:

X is the value of each observation.
$\bar{X}$ is the arithmetic mean of the values.
n is the number of observations in the sample.
$|\ |$ indicates the absolute value.

We take the absolute value of the deviations from the mean because if we didn't, the positive and negative deviations from the mean exactly offset each other, and the mean deviation would always be zero. Such a measure (zero) would be a useless statistic.

The mean deviation is computed by first determining the difference between each observation and the mean. These differences are then averaged without regard to their signs. For the PM statistics class the mean deviation is 4.0 years, found by the table on the right:

$X - \bar{X}$				Absolute Deviation
$\|17 - 21\|$	=	$\|-4\|$	=	4
$\|17 - 21\|$	=	$\|-4\|$	=	4
$\|18 - 21\|$	=	$\|-3\|$	=	3
$\|20 - 21\|$	=	$\|-1\|$	=	1
$\|25 - 21\|$	=	$\|-4\|$	=	4
$\|29 - 21\|$	=	$\|-8\|$	=	8
		Σ	=	24

Then

$$\bar{X} = \frac{\Sigma|X - \bar{X}|}{n} = \frac{24}{6} = 4$$

The parallel lines || indicate absolute value. To interpret, 4.0 years is the mean amount by which the ages differ from the arithmetic mean age of 21.0 years for the PM students.

The major characteristics of the mean deviation are:

1. All the observations are used in the calculations.
2. It is easy to understand.

The mean deviation has a disadvantage because of the use of the absolute values. Generally, absolute values are difficult to work with, so the mean deviation is not used as often as other measures of dispersion.

Variance and Standard Deviation

The disadvantage of the mean deviation is that the absolute values are difficult to manipulate mathematically. Squaring the differences from each value and the mean eliminates the problem of absolute values. These squared differences are used both in the computation of the ***variance*** and the ***standard deviation***.

> ***Variance:*** The arithmetic mean of the squared deviations from the mean.

The variance is non-negative and is zero only if all observations are the same.

> ***Standard Deviation:*** The square root of the variance

Squaring units of measurement, such as dollars or years, makes the variance cumbersome to use since it yields units like "dollars squared" or "years squared." However, by calculating the standard deviation, which is the positive square root of the variance, we can return to the original units, such as years or dollars. Because the standard deviation is easier to interpret, it is more widely used than the mean deviation or the variance.

Population Variance

The formula for the population variance and the sample variance are slightly different. The formula for the population variance is:

Population Variance $$\sigma^2 = \frac{\Sigma(X-\mu)^2}{N} \qquad [3-8]$$

Where:

σ^2 is the symbol for the population variance (σ is the Greek letter sigma). It is usually referred to as "sigma squared."

X is a value of an observation in the population.

μ is the arithmetic mean of the population.

N is the total number of observations in the population.

The major characteristics of the variance are:

3. All the observations are used in the calculations.
4. It is not influenced by extreme observations.
5. The units are somewhat difficult to work with. (They are the original units squared.)

Population Standard Deviation

The population standard deviation is the square root of the population variance. The formula for the population standard deviation is:

Population Standard Deviation $$\sigma = \sqrt{\frac{\Sigma(X-\mu)^2}{N}} \qquad [3-9]$$

Sample Variance

The conversion of the population variance formula to the sample variance formula is not as direct as the change made when we went from the population mean formula to the sample mean formula. Recall in that instance we replaced μ with $\overline{X}$ and N with n.

The conversion from population variance to sample variance requires a change in the denominator. Instead of substituting n, the number in the sample, for N, the number in the population, we replace N with $(n - 1)$. Thus the formula for the sample variance is:

Sample Variance

$$s^2 = \frac{\Sigma(X - \overline{X})^2}{n-1} \qquad [3-10]$$

Where:

s^2 is the symbol for the sample variance. It is pronounced as "*s* squared."
X is the value of each observation in the sample.
$\overline{X}$ is the mean of the sample.
n is the total number of observations in the sample.

Changing the denominator to $(n - 1)$ seems insignificant, however the use of n tends to underestimate the population variance. The use of $(n - 1)$ in the denominator provides an appropriate correction factor.

Sample Standard Deviation

The sample standard deviation is used as an estimator of the population standard deviation. The sample standard deviation is the square root of the sample variance. The formula is:

Standard Deviation

$$s = \sqrt{\frac{\Sigma(X - \overline{X})^2}{n-1}} \qquad [3-11]$$

Interpretation and Uses of the Standard Deviation

The standard deviation is used to measure the spread of the data. A small standard deviation indicates that the data is clustered close to the mean, thus the mean is representative of the data. A large standard deviation indicates that the data are spread out from the mean and the mean is not representative of the data.

Chebyshev's Theorem

We can use Chebyshev's theorem to determine the percent of the values that lie within a specified number of standard deviations of the mean.

> ***Chebyshev's theorem***: For any set of observations (sample or population), the proportion of the values that lie within k standard deviations of the mean is at least $1 - 1/k^2$, where k is any constant greater than 1.

The theorem holds for any set of observations regardless of the shape of the distribution.

The Empirical Rule

Chebyshev's theorem is concerned with any set of values: that is, the distribution of values can have any shape. If the distribution is approximately symmetrical and bell shaped, then the ***Empirical Rule*** or ***Normal Rule*** as it is often called is applied.

> ***Empirical Rule***: For a symmetrical, bell-shaped frequency distribution, approximately 68 percent of the observations will lie within plus and minus one standard deviation of the mean; about 95 percent of the observations will lie within plus and minus two standard deviations of the mean; and practically all (99.7 percent) will lie within plus and minus three standard deviations of the mean.

The rule states that:

- The mean, plus and minus one standard deviation, will include about 68% of the observations.
- The mean, plus and minus two standard deviations, will include about 95% of the observations.
- The mean, plus and minus three standard deviations, will include about 99.7% of the observations.

Glossary

Arithmetic mean: The sum of observations divided by the total number of observations.

Chebyshev's theorem: For any set of observations (sample or population), the minimum proportion of the values that lie within k standard deviations of the mean is at least $1 - 1/k^2$, where k is any constant greater than 1.

Empirical Rule: For a symmetrical, bell-shaped frequency distribution, approximately 68 percent of the observations will lie within plus and minus one standard deviation of the mean; about 95 percent of the observations will lie within plus and minus two standard deviations of the mean; and practically all (99.7 percent) will lie within plus and minus three standard deviations of the mean.

Geometric mean: The nth root of the product of n values.

Mean Deviation: The mean of the absolute values of the deviations from the arithmetic mean.

Measure of location: A single value that summarizes a set of data. It locates the center of the values.

Median: The midpoints of the values after all observations have been ordered from the smallest to the largest, or from largest to smallest. Fifty percent of the observations are above the median and 50 percent are below the median.

Mode: The value of the observation that appears most frequently.

Negatively skewed distribution: The long tail is to the left or in the negative direction. The mean is smaller than the median or mode.

Parameter: A characteristic of a population.

Positively skewed distribution: The long tail is to the right; that is, in the positive direction. The mean is larger than the median or the mode.

Range: The difference between the largest and smallest values in a data set.

Standard Deviation: The square root of the variance.

Statistic: A characteristic of a sample.

Weighted mean: The value of each observation is multiplied by the number of times it occurs. The sum of these products is divided by the total number of observations to determine the weighted mean.

Symmetrical distribution: A distribution that has the same shape on either side of the median.

Variance: The arithmetic mean of the squared deviations from the mean.

Chapter Examples

Example 1 – Finding Mean, Median and Mode

A comparison shopper employed by a large grocery chain recorded these prices for a 340-gram jar of Kraft blackberry preserves at a sample of six supermarkets selected at random.

Supermarket	Price X
1	$1.31
2	1.35
3	1.26
4	1.42
5	1.31
6	1.33
Total	$7.98

a. Compute the arithmetic mean.

b. Compute the median.

c. Compute the mode.

Solution 1 – Finding Mean, Median and Mode

a. Determine the mean price of this raw data by summing the prices for the six jars and dividing the total by six. Recall the formula for the mean of a sample was given previously. See formula [3-2].

$$\overline{X} = \frac{\Sigma X}{n} = \frac{\$7.98}{6} = \$1.33$$

b. As noted above the *median* is defined as the middle value of a set of data, after the data is arranged from smallest to largest. The prices for the six jars of blackberry preserves have been ordered from a low of $1.26 up to $1.42. Because this is an even number of prices the median price is halfway between the third and the fourth price. The median is $1.32.

Prices Arranged from Low to High:

$1.26 $1.31 $1.31 $1.33 $1.35 $1.42

⇑ ⇑

$$\text{Median} = \frac{\$1.31 + \$1.33}{2} = \$1.32$$

Suppose there are an odd number of blackberry preserve prices, such as shown in the table.

$1.31	$1.31	$1.33	$1.35	$1.42

The median is the middle value ($1.33). To find the median, the values must first be ordered from low to high.

c. The mode is the price that occurs most often. The price of \$1.31 occurs twice in the original data and is the mode.

Self–Review 3.1

Check your answers against those in the ANSWER section.

The number of semester credit hours for seven part-time college students is: 8, 5, 4, 10, 8, 3, and 4. Compute the:

a. mean b. median c. mode

Example 2 – Finding the Geometric Mean

From 1990 to 2004 the number of cell phones sold per month by Wagoner Enterprises increased from 5 to 300 (in thousands). Compute the mean annual percent increase in the number of cell phones sold.

Solution 2 – Finding the Geometric Mean

The geometric mean (*GM*) annual percent increase from one time period to another is determined using formula [3-5].

$$GM = \sqrt[n]{\frac{\text{Value at the end of the period}}{\text{Value at the start of the period}}} - 1 \qquad [3\text{-}5]$$

Note that there are 14 years between 1990 and 2004, so, $n = 14$.

$$GM = \sqrt[14]{\frac{300}{5}} - 1 = \sqrt[14]{60.0} - 1 = 1.33971 - 1.00000 = 0.33971$$

For those with a $\sqrt[x]{y}$ key on their calculator, the geometric mean can be solved quickly by:

$$GM = \sqrt[n]{\frac{300}{5}} - 1 = \sqrt[14]{60.0} - 1$$

Using $\sqrt[x]{y}$	Display
300 ÷ 5 =	60
Depress $\sqrt[x]{y}$	
Depress 14	1.33971
Depress −1 =	0.33971, or about 34%

The value 1 is subtracted, according to formula [3-5], so the rate of increase is 0.33971, or 33.971% per year. The sale of hospital beds increased at a rate of almost 34% per year.

Self–Review 3.2

Check your answers against those in the ANSWER section.

In 1989, thirty acres of woods was valued at \$475 per acre. In 2004 the acreage was valued at \$2850 per acre. What is the geometric mean annual percent increase in value?

Example 3 – Finding the Range

A sample of the amounts spent in November for propane gas to heat homes of similar sizes in Duluth revealed these amounts (to the nearest dollar):

$191 $212 $176 $129 $106 $92 $108 $109 $103 $121 $175 $194

What is the range? Interpret your results.

Solution 3 – Finding the Range

Recall that the range is the difference between the largest value and the smallest value.

$$\text{Range} = \text{Largest Value} - \text{Smallest Value} = (\$212 - \$92) = \$120$$

This indicates that there is a difference of $120 between the largest and the smallest heating cost.

Example 4 – Finding the Mean Deviation

Using the heating cost data in Example 3, compute the mean deviation.

Solution 4 – Finding the Mean Deviation

The mean deviation is the mean of the absolute deviations from the arithmetic mean. For raw, or ungrouped data, it is computed by first determining the mean. Next, find the difference between each value and the arithmetic mean. Finally, total these differences and divide the total by the number of observations. We ignore the sign of each difference. Formula [3-2] for the sample mean and formula [3-7] for the mean deviation are shown below.

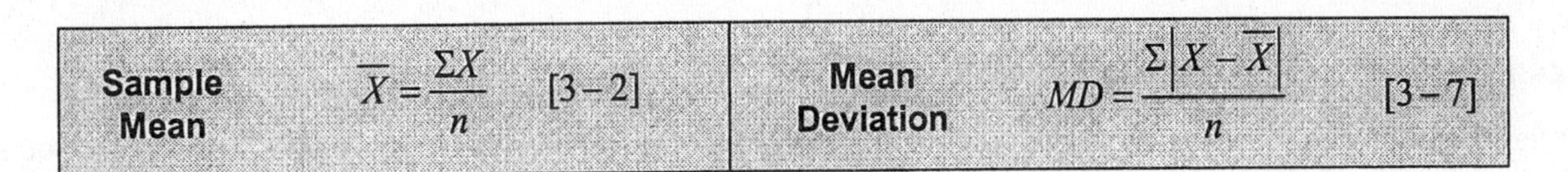

Sample Mean	$\overline{X} = \frac{\Sigma X}{n}$	[3-2]	Mean Deviation	$MD = \frac{\Sigma\lvert X - \overline{X}\rvert}{n}$	[3-7]

The table below shows the data values, each data value minus the mean, and the absolute value of the deviations from the mean.

In other words, the signs of the deviations from the mean are disregarded.

Payment X	$\lvert X-\overline{X}\rvert$		Absolute Deviations
$191	|$+48|	=	$48
212	|+69|	=	69
176	|+33|	=	33
129	|−14|	=	14
106	|−37|	=	37
92	|−51|	=	51
108	|−35|	=	35
109	|−34|	=	34
103	|−40|	=	40
121	|−22|	=	22
175	|+32|	=	32
194	|+51|	=	51
$1,716			$466

$$\overline{X} = \frac{\Sigma X}{n} = \frac{\$1,716}{12} = \$143.00$$

$$MD = \frac{\Sigma\left|X-\overline{X}\right|}{n} = \frac{\$466}{12} = \$38.83$$

The mean deviation indicates that the typical electric bill deviates $38.83 from the mean of $143.00.

Self–Review 3.3

Check your answers against those in the ANSWER section.

A sample of the amount of rent paid for one bedroom apartments of similar size near the University of Akron are:

$295 $475 $345 $595 $538 $460 $495 $422 $370 $333 $370 $390

a. What is the range? Interpret your results. **b.** Compute the mean deviation.

Example 5 – Finding the Weighted Mean

At Sarasota College there are 10 instructors, 12 assistant professors, 20 associate professors, and 5 professors. Their average annual salaries are $34,000, $45,000, $58,000, and $68,000, respectively. What is the weighted mean salary?

Solution 5 – Finding the Weighted Mean

The number of faculty for each rank is not equal. Therefore, it is not appropriate simply to add the average salaries of the four ranks and divide by 4. We have a better method for weighting the averages. In this problem the salaries for each rank are multiplied by the number of faculty in that rank, the products totaled, then divided by the number of faculty. The result is the weighted mean.

$$\overline{X} = \frac{w_1X_1 + w_2X_2 + w_3X_3 + w_4X_4}{w_1 + w_2 + w_{31} + w_4}$$

$$= \frac{10(\$34,000) + 12(\$45,000) + 20(\$58,000) + 5(\$68,000)}{10 + 12 + 20 + 5}$$

$$= \frac{\$2,380,000}{47}$$

$$= \$50,638$$

Self–Review 3.4

Check your answers against those in the ANSWER section.

During the past month an electronics store sold 31 model EL733 calculators for $30 each, 42 model EL480 calculators for $10 each, 47 model FX115 calculators for $20 each, and 63 model BA35 calculators for $24 each.

What is the weighted mean price of the calculators?

Example 6 – Finding the Variance and the Standard Deviation

Using the same heating cost data in Example 3, compute the variance and the standard deviation.

Solution 6 – Finding the Variance and the Standard Deviation

The sample variance, designated s^2, is based on squared deviations from the mean. For ungrouped raw data, it is computed using formula [3-10].

$$s^2 = \frac{\Sigma(X-\overline{X})^2}{n-1} \qquad [3-10]$$

To compute the variance:

Step 1: Compute the mean.

Step 2: Find the difference between each observation and the mean. Square the difference.

Step 3: Find the sum of all the squared differences.

Step 4: Divide the sum of the squared differences by the number of items in the sample minus one.

X	$\overline{X}$	$(X-\overline{X})$	$(X-\overline{X})^2$
$191	$143	$48	2,304
212	143	69	4,761
176	143	33	1,089
129	143	–14	196
106	143	–37	1,369
92	143	–51	2,601
108	143	–35	1,225
109	143	–34	1,156
103	143	–40	1,600
121	143	–22	484
175	143	32	1,024
194	143	51	2,601
$1,716	0	0	20,410

$$\overline{X} = \frac{\Sigma X}{n} = \frac{1{,}716}{12} = 143$$

$$s^2 = \frac{\Sigma(X-\overline{X})^2}{n-1} = \frac{20{,}410}{12-1} = 1{,}855.45$$

$$s = \sqrt{\frac{\Sigma(X-\overline{X})^2}{n-1}} = \sqrt{1{,}855.45} = 43.074 = 43.07$$

The standard deviation of the sample, designated by s, is the square root of the variance. The square root of 1,855.45 is \$43.07. Note that the standard deviation is in the same unit as the original data, that is, dollars.

Self–Review 3.5

Check your answers against those in the ANSWER section.

The manager of a fast-food restaurant selected several cash register receipts at random. The amounts spent by customers were \$12, \$15, \$16, \$10, and \$27. Compute the:

a. range **b.** the mean **c.** the sample variance **d.** the sample standard deviation

Example 7 – Using Chebyshev's Theorem and the Empirical Rule

A sample of the business faculty at state-supported institutions in Ohio revealed the mean income to be \$52,000 for 9 months with a standard deviation of \$3,000. Use Chebyshev's Theorem and the Empirical Rule to estimate the proportion of faculty who earn more than \$46,000 but less than \$58,000.

Solution 7 – Using Chebyshev's Theorem and the Empirical Rule

To find the proportion of faculty who earn between \$46,000 and \$58,000 we must first determine k; k is the number of standard deviations above or below the mean.

$$k = \frac{X - \overline{X}}{s} = \frac{\$46{,}000 - \$52{,}000}{\$3{,}000} = -2.00$$

$$k = \frac{X - \overline{X}}{s} = \frac{\$58{,}000 - \$52{,}000}{\$3{,}000} = 2.00$$

Applying Chebyshev's theorem: $1 - \frac{1}{k^2} = 1 - \frac{1}{2^2} = 0.75$

This means that at least 75 percent of the faculty earn between \$46,000 and \$58,000.

The Empirical rule states that about 68 percent of the observations fall within one standard deviation of the mean, 95 percent are within plus and minus two standard deviations of the mean, and virtually all (99.7%) will lie within three standard deviations from the mean. Hence, about 95 percent of the observations fall between \$46,000 and \$58,000, found by $\overline{X} \pm 2s = \$52{,}000 \pm 2(\$3{,}000)$. If we conclude that we have a bell shaped distribution, most of the observations fall within the interval.

Self–Review 3.6

Check your answers against those in the ANSWER section.

An automobile dealership pays its salespeople a salary plus a commission on sales. The mean biweekly commission is \$990, the median \$950, and the standard deviation \$70.

a. Estimate the percent of salespeople that earn more than \$885, but less than \$1095.
b. Is the distribution of commissions positively skewed, negatively skewed, or symmetrical?

CHAPTER 3 ASSIGNMENT

DESCRIBING DATA: NUMERICAL MEASURES

Name ______________________________ Section ____________ Score________

Part I Select the correct answer and write the appropriate letter in the space provided.

_____ **1.** The arithmetic mean is computed by
- **a.** finding the value that occurs most often.
- **b.** finding the middle observation and dividing by 2.
- **c.** summing the values and dividing by the number of values.
- **d.** electing the value in the middle of the data set.

_____ **2.** To compute the arithmetic mean at least the
- **a.** nominal level of measurement is required.
- **b.** ordinal level of measurement is required.
- **c.** interval level of measurement is required.
- **d.** ratio level of measurement is required.

_____ **3.** The value that occurs most often in a set of data is called the
- **a.** mean.
- **b.** median.
- **c.** geometric mean.
- **d.** mode.

_____ **4.** What level of measurement is required to determine the mode?
- **a.** nominal
- **b.** ordinal
- **c.** interval
- **d.** ratio

_____ **5.** For a symmetric distribution
- **a.** the mean is larger than the median.
- **b.** the mode is the largest value.
- **c.** the mean is smaller than the median.
- **d.** the mean, median and the mode are equal.

_____ **6.** Which of the following is ***not*** true about the arithmetic mean.
- **a.** all the values are used in its calculation
- **b.** half of the observations are always larger than the mean
- **c.** it is influenced by a large value
- **d.** it is found by summing all the values and dividing by the number of observations

_____ **7.** In a *negatively* skewed distribution
- **a.** the mean is smaller than the median.
- **b.** the mean is larger than the median.
- **c.** the mean and median are equal.
- **d.** the median and the mode are equal.

______ 8. What level of measurement is required for the median?

a. nominal b. ordinal
c. interval d. ratio

______ 9. The Dow Jones Industrial Average increased from 961 in 1980 to over 9,500 in the third quarter of 2003. The annual rate of increase is best described by the

a. geometric mean. b. weighted mean.
c. median. d. mode.

______ 10. What is the shape of a frequency distribution with an arithmetic mean of 12,000 pounds, a median of 12,000 pounds, and a mode of 12,000 pounds?

a. flat b. symmetric
c. geometrically skewed d. positively skewed

______ 11. The mean deviation

a. is the average of all the values.
b. is the midpoint of the range.
c. is the average of how far each value is from the median.
d. is the average of how far each value is from the mean.

______ 12. The sum of the deviations from the mean is always

a. equal to the mean.
b. equal to zero.
c. always positive.
d. equal to the median.

______ 13. The square of the standard deviation is equal to

a. the mean.
b. the variance.
c. the median.
d. the mean deviation.

______ 14. What is the shape of a frequency distribution with an arithmetic mean of 800 pounds, median of 758 pounds, and a mode of 750 pounds?

a. negatively skewed b. symmetric
c. geometrically skewed d. positively skewed

______ 15. According to the Empirical Rule, about what percent of the observations are within 2 standard deviations of the mean?

a. 50 b. 68
c. 99.7 d. 95

Part II Find the answers to each of the following questions. Show essential calculations.

16. A study conducted by the Toledo police at the intersection of Byrne and Heatherdowns for the 7 to 9 AM drive time revealed the following number of vehicles proceeded through the intersection after the light changed. The information reported below is for a sample of seven days during a six month period

6 12 7 12 8 4 5

a. Compute the range.

a.

b. Compute the sample mean.

b.

c. What is the median?

c.

d. What is the mode?

d.

e. Describe the skewness.

e.

17. A shipment of packages to the Solomon Company included 10 packages weighing 7.4 pounds, 12 weighing 8.2 pounds and 6 weighing 8.7 pounds. What value would you use as a typical amount for the weight of a package?

17.

18. The mean daily attendance for eight large employers in Dade County is as follows:

95.7%, 95.3%, 95.5%, 95%, 94.7%, 93.7%, 93.8% 90.7%.

Find the average daily attendance for the county.

18.

19. From 1983 to 2004 the net sales for the J.M. Smucker Company increased from $157 million to $1,311 million. Compute the mean annual percent increase in net sales.

19.

20. The salaries of the eleven employees at Marker Graphics are given (in thousands).

15 17 23 26 27 35 72 88 91 98 102

a. Compute the range.

a.

b. Compute the mean deviation.

b.

c.

c. Compute the standard deviation.

21. The mean number of gallons of gasoline pumped per customer at Ray's Marathon Station is 9.5 gallons with a standard deviation of 0.75 gallons. The median number of gallons pumped is 10.0 gallons. The arithmetic mean amount of time spent by a customer in the station is 6.5 minutes with a standard deviation of 2 minutes.

a. According to Chebyshev's Theorem, what proportion (percent) of the customers spend between 3.30 minutes and 9.70 minutes at the station?

a.

b. According to the Empirical Rule, what proportion of the customers pump between 8.00 gallons and 11.00 gallons?

b.

CHAPTER 4
DESCRIBING DATA:
DISPLAYING AND EXPLORING DATA

Chapter Goals

When you have completed this chapter, you will be able to:

1. Develop and interpret a ***dot plot***.
2. Develop and interpret ***quartiles***, ***deciles***, and ***percentiles***.
3. Construct and interpret ***box plots***.
4. Compute and understand the ***coefficient of skewness***.
5. Draw and interpret a ***scatter diagram***.
6. Set up and interpret a ***contingency table***.

Introduction

We continue our study of descriptive statistics with measures of dispersion, such as dot plots, quartiles, percentiles, and box plots. Dot plots and box plots give additional insight into where the values are concentrated and dispersed and the general shape of the data. Finally we consider bivariate data where we observe two variables for each individual or observation selected.

Dot Plots

In Chapter 2 we grouped data in classes and constructed a histogram. When we organize the data into classes, we lose the exact value of the observations. ***Dot plots*** group data as little as possible, hence we do not lose the identity of the individual observations.

> ***Dot Plot***: A graph for displaying a set of data. Each numerical value is represented by a dot placed above a horizontal number line.

To develop a dot plot we display a dot for each observation along a horizontal number line indicating the value of each piece of data. For multiple observations we pile the dots on top of each other.

The steps to follow in developing a dot plot graph are:

1. Sort the data from smallest to largest.
2. Draw and label a number line.
3. Place a dot ● for each observation.

Length of Service (in years)					
7	6	2	10	6	6
5	8	4	8	4	7
6	5	3	3	7	5

As an example, the lengths of service, in years, of a sample of eighteen employees are given.

Step 1: Sort the data from smallest to largest.

2	3	3	4	4	5	5	5	6	6	6	6	7	7	7	8	8	10

Step 2: Draw the number line and label it as shown.

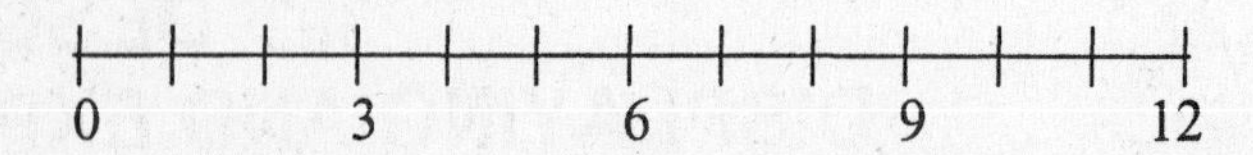

Step 3: Place a dot ● for each observation.

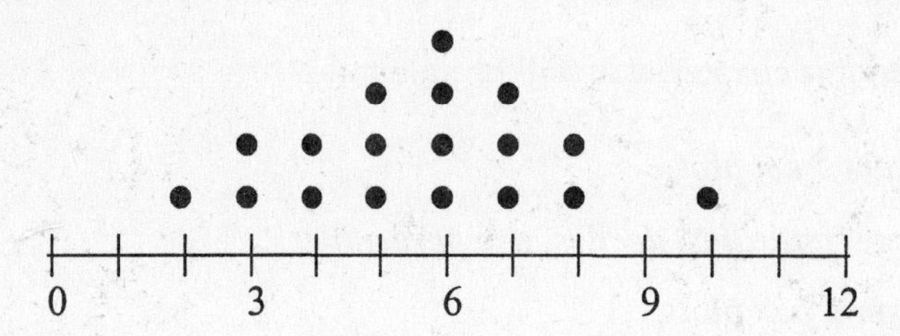

We note that the data range is from 2 to 10 years and that the data clusters around 6 years.

Other Measures of Dispersion

The standard deviation is the most widely used measure of dispersion. However there are several others, which include ***Quartiles***, ***Deciles***, and ***Percentiles***.

Quartiles

Recall that the median divides data that has been placed in order from smallest to largest, such that half the values are below the median and half are above the median. If we divide the lower and upper set of values into two equal parts, we have quartiles. Quartiles divide a set of data into four equal parts.

> ***First Quartile*** The point below which one-fourth or 25% of the ranked data values lie. (It is designated Q_1)

> ***Third Quartile*** The point below which three-fourths or 75% of the ranked data values lie. (It is designated Q_3)

Logically the median is the ***Second Quartile*** (designated Q_2). The values corresponding to Q_1 , Q_2 *and* Q_3 divide a set of data into four equal parts.

Deciles and Percentiles

Just as quartiles divide a distribution into 4 equal parts, deciles divide a distribution into ten equal parts; and percentiles divide a distribution into 100 equal parts.

For example: If you were told that your Scholastic Aptitude Test score was in the 9th decile, you could assume that 90 percent of those taking the test had a lower score than yours and that 10 percent had a higher score. A grade point average in the 55th percentile means that 55 percent of students have a lower GPA than yours and that 45 percent have a higher GPA.

The procedure for finding the quartile, decile, and a percentile for ungrouped data is to order the data from smallest to largest. Then use text formula [4-1].

Location of a Percentile	$L_p = (n+1)\frac{P}{100}$	[4–1]

Where:

L_p refers to the location of the desired percentile.
n is the number of observations.
P is the desired percentile

Note that this is a generic formula for percentiles, deciles and quartiles.

For example, if you had a set of data with 49 observations in ordered array and wanted to locate the 78th percentile, then let $P = 78$ and $n = 49$ so $L_p = (n+1)\frac{P}{100} = (49+1)\frac{78}{100} = 39$. Thus you would locate the 39th observation.

If you wanted to locate the 6th decile, first note that the 6th decile equals the 60 percentile. Then let $P = 60$ and $L_p = (n+1)\frac{P}{100} = (49+1)\frac{60}{100} = 30$. Thus you would locate the 30th observation.

Box Plots

A ***box plot*** is a graphical display that helps us picture how a set of data is distributed relative to the quartiles.

> ***Box plot***: A graphical display based on five statistics: the minimum value, Q_1 (the first quartile), Q_2 the median, Q_3 (the third quartile) and the maximum value.

To construct a box plot we need five pieces of information. We need the minimum value, Q_1 (the first quartile), Q_2 the median, Q_3 (the third quartile) and the maximum value. The details for constructing and interpreting a box plot are found in Example 3 of the Chapter Examples.

Skewness

Another characteristic of a set of data is the shape of the distribution. There are four shapes commonly observed: **symmetric, positively skewed, negatively skewed**, and **bimodal**. The measures of location and the measures of dispersion are both descriptive characteristics of a set of data.

A third characteristic of a distribution is its ***skewness***. As noted before, a **symmetric** distribution has the same shape on either side of the median and it has no skewness. For a **positively skewed** distribution the long tail is to the right, the mean is larger than the median or the mode, and the mode appears at the highest point on the curve. For a **negatively skewed** distribution the mode is the largest value and is at the highest point of the curve, while the mean is the smallest. A bimodal distribution will have two or more peaks. The ***coefficient of skewness*** is used to describe how a distribution is skewed.

> ***Coefficient of skewness:*** A measure to describe the degree of skewness.

Text formula [4–2] is for Pearson's Coefficient of Skewness.

Pearson's Coefficient of Skewness	$sk = \frac{3(\bar{X} - \text{Median})}{s}$	[4–2]

Where:

sk is the coefficient of skewness.
$\bar{X}$ is the mean.
s is the standard deviation.

Characteristics of the coefficient of skewness are:

- The coefficient of skewness, designated sk, measures the amount of skewness and may range from –3.0 to +3.0.
- A value near –3, such as –2.57, indicates considerable negative skewness.
- A value such as 1.63 indicates moderate positive skewness.
- A value of 0, which will occur when the mean and median are equal, indicates the distribution is symmetrical and that there is no skewness.

This information is summarized in the chart.

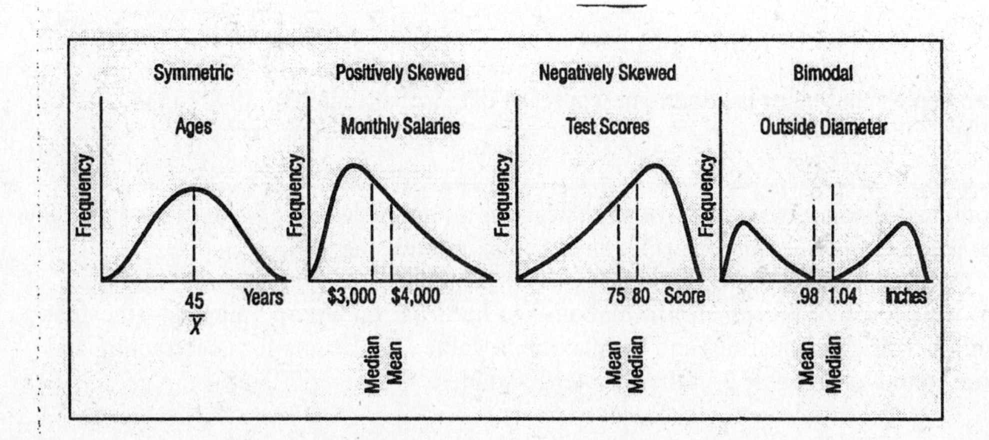

Describing the Relationship Between Two Variables

To summarize the distribution of a set of data, in Chapter 2 we used a histogram and in the first part of this chapter we used dot plots. We studied a single variable sometimes called univariate data.

When we study the relationship between two variables we refer to the data as ***bivariate***.

Bivariate data: A collection of paired data values.

For example a realtor might want to study the relationship between the selling price of a home and the number of days the home is on the market.

One graphical technique used to show the relationship between variables is a ***scatter diagram***.

> ***Scatter diagram:*** A graph in which paired data values are plotted on an *x,y Axis*.

The steps to follow in developing a scatter diagram are:

1. We need two variables.
2. We scale one variable (x) along the horizontal axis (x – *Axis*) of a graph and the corresponding variable (y) along the vertical axis (y – *Axis*).
3. Place a dot ● for each (x, y) pair of observations.

Usually one variable depends on another.

The scatter diagram shows the relationship between the location by zip code and the selling price per square foot for 17 properties of 4 units or more which recently sold in the 436__ Zip code area.

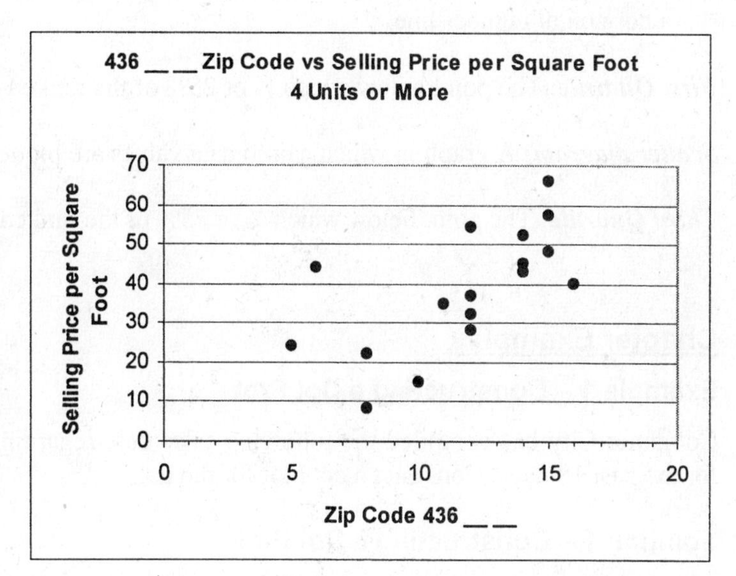

It appears that as the last two digits of the Zip code increase, the price per square foot of the apartment building increases.

A real estate sales agent suggested that the lower the price per square foot, the lower the last two digits of the zip code number. In other words as the last two digits of the Zip code number increased so did the price per square foot.

Contingency Table

When we study the relationship between two or more variables when one or both are nominal or ratio scale, we tally the results into a two-way table. This two-way table is referred to as a ***contingency table.***

> ***Contingency table:*** A table used to classify sample observations according to two identifiable characteristics.

A contingency table is a cross tabulation that simultaneously summarizes two variables of interest and their relationship.

A survey of 60 school children classified each as to gender and the number of times lunch was purchased at school during a four-week period. Each respondent is classified according to two criteria – the number of times lunch was purchased and gender.

	Gender		
Bought Lunch	**Boys**	**Girls**	**Total**
0 up to 10	10	5	15
10 up to 20	20	25	45
Total	30	30	60

Glossary

Bivariate data: A collection of paired data values.

Box Plot: A graphical display based on five statistics: the minimum value, Q_1 (the first quartile), Q_2 the median, Q_3 (the third quartile) and the maximum value.

Coefficient of skewness: A measure to describe the degree of skewness.

Contingency table: A table used to classify sample observations according to two identifiable characteristics.

Dot Plot: A graph for displaying a set of data. Each numerical value is represented by a dot placed above a horizontal number line.

First Quartile: The point below which ¼ or 25% of the ranked data values lie. (It is designated Q_1).

Scatter diagram: A graph in which paired data values are plotted on an *x,y Axis*

Third Quartile: The point below which ¾ or 75% of the ranked data values lie. (It is designated Q_3)

Chapter Examples

Example 1 – Constructing a Dot Plot

Computer City has compiled the following sales data regarding the number of computers sold each day for the past 18 days. Construct a dot plot for the data.

Solution 1 – Constructing a Dot Plot

Step 1: Sort the data from smallest to largest.

Number of computers sold					
8	12	5	12	16	9
11	18	9	11	12	10
11	15	3	10	10	12

3	5	8	9	9	10	10	10	11	11	11	12	12	12	12	15	16	18

Step 2: Draw the number line and label it as shown.

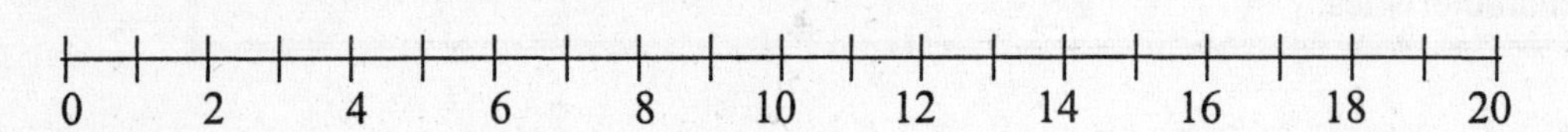

Step 3: Place a dot ● for each observation.

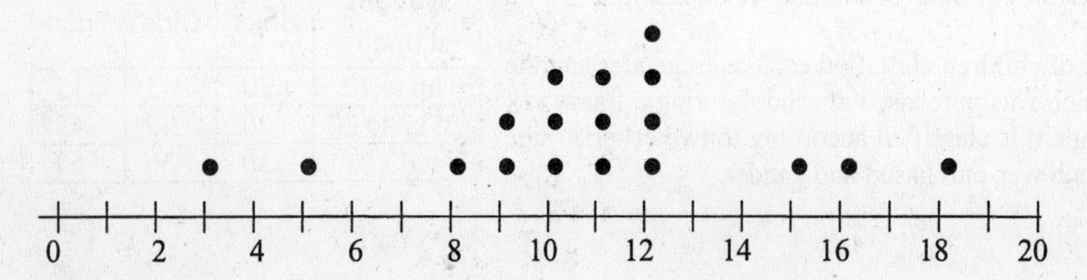

We note the data range is from 3 to 18 computers and the data clusters around 11 computers.

Self-Review 4.1

Check your answers against those in the ANSWER section.

Listed below are the selling prices (in thousands of dollars) of a sample of 20 vehicles sold by salespeople employed by Higginbotham Auto Group in Hancock County, Ohio.

26	21	18	22	29
28	28	25	28	24
30	22	35	35	25
35	25	20	37	26

a. Sort the data from low to high. **b.** Draw a dot plot for the data.

Example 2 – Calculating Quartiles

The selling prices (in thousands of dollars) of a sample of 15 homes sold by agents working in Lucas County, Ohio are given. Determine the following:

\$20	\$56	\$65	\$17	\$26	\$90	\$13	\$27
\$16	\$68	\$86	\$80	\$50	\$25	\$92	

a. The first quartile. **b.** The third quartile. **c.** Determine the median.

Solution 2 – Calculating Quartiles

Step 1: Organize the 15 observations into an ordered array from smallest to largest:

13	16	17	20	25	26	27	50	56	65	68	80	86	90	92

Step 2: Locate the first quartile, let $P = 25$ and $L_p = (n+1)\frac{P}{100} = (15+1)\frac{25}{100} = 4$

a. Step 3: Locate the 4th observation in the array which is 20. Thus $Q_1 = 20$ or \$20,000.

Step 4: Locate the third quartile, let $P = 75$ and $L_p = (n+1)\frac{P}{100} = (15+1)\frac{75}{100} = 12$

b. Step 5: Locate the 12th observation in the array which is 80. Thus $Q_3 = 80$ or \$80,000.

Step 6: Locate the median, let $P = 50$ and $L_p = (n+1)\frac{P}{100} = (15+1)\frac{50}{100} = 8$

c. Step 7: Locate the 8th observation in the array which is 50. Thus Q_2 = the median =50 or \$50,000.

In the above example with 15 observations the location formula yielded a whole number result. Suppose we were to add one more observation (95) to the data list.

13	16	17	20	25	26	27	50	56	65	68	80	86	90	92	95

What is the third quartile now?

To locate the third quartile, let $P = 75$ and $n = 16$, so $L_p = (n+1)\frac{P}{100} = (16+1)\frac{75}{100} = 12.75$

Then locate the 12th and 13th observation in the array which are 80 and 86. The value of the third quartile is 0.75 of the distance between the 12th and 13th value. We must calculate 0.75(86 – 80) = 4.5 Thus Q_3 = (80 + 4.5) = 84.5 or $84,500.

Example 3 – Developing a Box Plot

Use the selling price of homes data from Example 2 to develop a box plot.

Solution 3 – Developing a Box Plot

Step 1: Identify the five essential pieces of data:

Minimum value = 13, $Q_1 = 20$, $Q_2 = 50$ $Q_3 = 80$, Maximum value = 92

Step 2: Create an appropriate scale along the horizontal axis.

Step 3: Draw a box that starts at $Q_1 = 20$, and ends at $Q_3 = 80$. Inside the box we place a vertical line to represent the median 50. We then extend horizontal lines from the box to the minimum (12) and the maximum (92).

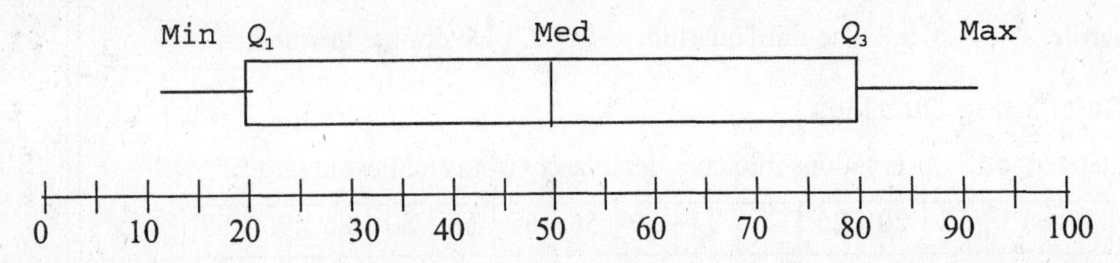

The box plot shows that the middle 50 percent of the homes sold for between $20,000 and $80,000. Also the distribution is somewhat positively skewed, since the line from Q_3 (80) to the maximum (92) is longer than the line from Q_1 (20) to the minimum (13).

In other words the 25% of the data larger than the third quartile is spread out more than the 25% of the data less than the first quartile.

Self-Review 4.2

Check your answers against those in the ANSWER section.

Listed below are the selling prices (in thousands of dollars) of a sample of 19 homes sold by agents working in Franklin County, Ohio.

86	61	148	81
39	142	140	65
28	85	90	92
25	50	85	85
82	120	137	

Determine the following:

a. the first quartile **b.** the third quartile
c. the median **d.** Draw a box plot for the data.

Example 4 – Determining the Coefficient of Skewness

The research director of a large oil company conducted a study of the buying habits of consumers with respect to the amount of gasoline purchased with credit cards at the pump. The arithmetic mean amount is 11.50 gallons, and the median amount is 11.95 gallons. The standard deviation of the sample is 4.5 gallons. Determine the coefficient of skewness. Comment on the shape of the distribution.

Solution 4 – Determining the Coefficient of Skewness

The coefficient of skewness measures the general shape of the distribution. A distribution that is symmetrical has no skewness and the coefficient of skewness is 0. Skewness ranges from –3 to +3. The direction of the long tail of the distribution points in the direction of the skewness. If the mean is larger than the median, the skewness is positive. If the median is larger than the mean, the skewness is negative.

The coefficient of skewness is found by formula [4-2]. For this problem:

$$sk = \frac{3(\overline{X} - \text{median})}{s}$$
$$= \frac{3(11.50 - 11.95)}{4.5}$$
$$= -0.30$$

This indicates that there is a slight negative skewness in the distribution of gasoline purchases with credit cards at the pump.

Self-Review 4.3

Check your answers against those in the ANSWER section.

An automobile dealership pays its salespeople a salary plus a commission on sales. The mean biweekly commission is $1385, the median $1330, and the standard deviation $75.

a. Is the distribution of commissions positively skewed, negatively skewed, or symmetrical?
b. Compute the coefficient of skewness to verify your answer.

Example 5 – Constructing a Scatter Diagram

It is believed that the annual repair cost for a commercial dishwasher is related to its age. A sample of 10 dishwashers resulted in the scatter diagram shown at the right.

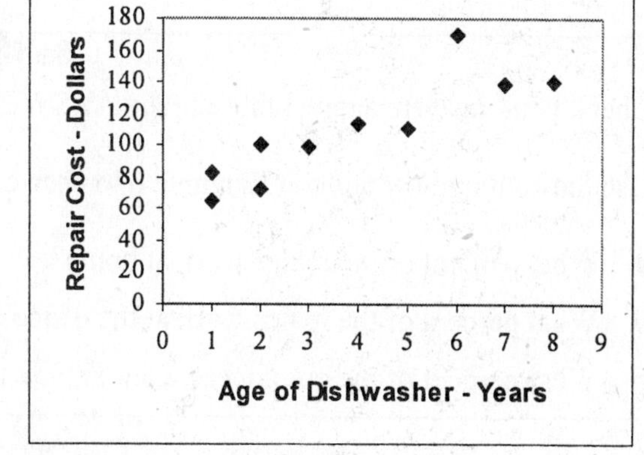

a. How many dishwashers were studied?

b. Estimate the repair cost for a 7-year old dishwasher.

c. How would you characterize the relationship between age and repair cost?

Solution 5 – Constructing a Scatter Diagram

a. Counting the points reveals that 10 dishwashers were studied.

b. The estimated repair cost for a 7-year old dishwasher is approximately $140.

c. It appears that the older the dishwasher, the higher the repair cost.

Self-Review 4.4

Check your answers against those in the ANSWER section.

The fuel tank capacity in gallons and the cruising range is given for 6 SUV's.

a. Develop a scatter diagram for the data.

b. How would you characterize the relationship between fuel tank capacity and cruising range?

Capacity in gallons	Cruising range in miles
22	445
21	405
23	410
24	420
24	365
25	380

Example 6 – Using the Contingency Table

A survey of the 40 employees of Systems Consultants LLC classified each as to gender and how many days a month they spent working at home or at the company office place.

Work	Gender		
Place	Male	Female	Total
Home	8	14	22
Office	12	6	18
Total	20	20	40

a. What percent of the females work at home?

b. What percent of the employees work at home?

c. How would you characterize the relationship between working at home and gender?

Solution 6 – Using the Contingency Table

a. Percent females working at home $= \frac{14}{20} = 0.70 = 70\%$

b. Percent employees working at home $= \frac{22}{40} = 0.55 = 55\%$

c. Obviously more females prefer to work at home

Self-Review 4.5

Check your answers against those in the ANSWER section.

Use the contingency table in Example 6 to answer the following:

a. What percent of the males work at home?

b. What percent of the males work at the office?

c. What percent of the employees work at the office?

CHAPTER 4 ASSIGNMENT

DESCRIBING DATA: DISPLAYING AND EXPLORING DATA

Name ________________________ Section ______________ Score __________

Part I Select the correct answer and write the appropriate letter in the space provided.

______ **1.** A dot plot shows the range of values along the:
a. vertical axis. **b.** (x, y) diagonal axis.
c. median axis. **d.** horizontal axis.

______ **2.** When displaying data with a dot plot, we:
a. do not lose the identity of an individual data point
b. are able to show the range of the values.
c. are able to see the shape of the distribution.
d. all of the above are correct.

______ **3.** A scatter diagram is a graphic tool used to portray:
a. the mean of the data values. **b.** the range of the data values.
c. the midpoint of data values. **d.** the shape of the distribution.

______ **4.** Which one of the following measures of dispersion does *not* divide a set of observations into equal parts?
a. quartiles **b.** deciles
c. percentiles **d.** standard deviations

______ **5.** The inter quartile range is the difference between:
a. the second and third quartile. **b.** the second and fourth quartile.
c. the first and third quartile. **d.** the first and last quartile.

______ **6.** A box plot is based on:
a. the mean.
b. percentiles.
c. deciles.
d. the first and third quartile, the median, the maximum and the minimum.

______ **7.** The interquartile range is based on:
a. the median.
b. the mean deviation.
c. the square of the mean deviation.
d. based on the middle 50 percent of the observations.

______ **8.** The coefficient of skewness is a measure:
a. of the relationship of the mode and median.
b. based on the mean deviation.
c. of the symmetry of a distribution.
d. based on the middle 50 percent of the observations.

Part II Show all of your work. Write the answer in the space provided.

9. A sample of 30 homes sold during the past year by Gomminger Realty Company was selected for study. (Selling price is reported in thousands of dollars.)

76	74	71	78	80	67
80	82	67	88	72	78
85	76	84	82	83	80
72	82	77	79	86	89
69	70	82	86	78	77

a. Construct a dot plot for the data.

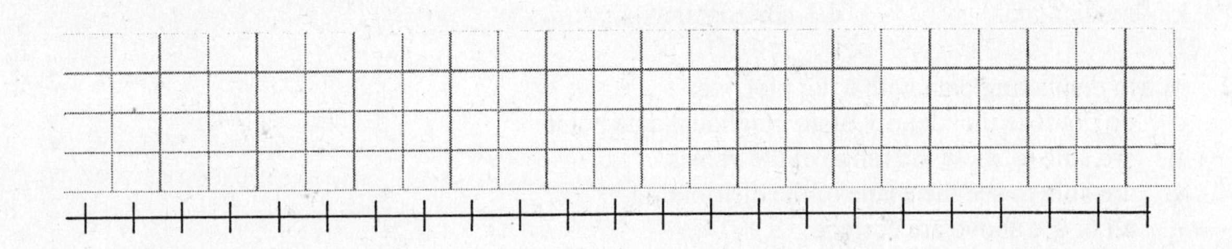

10. The revenues of the top eleven personal computer manufacturers are given (in hundred millions).

15	17	23	26	27	35
72	88	91	98	102	

a. Compute the first quartile.

a. Q_1
b. $Q3$
c. M

b. The third quartile.

c. Determine the median.

d. Draw a box plot for the data.

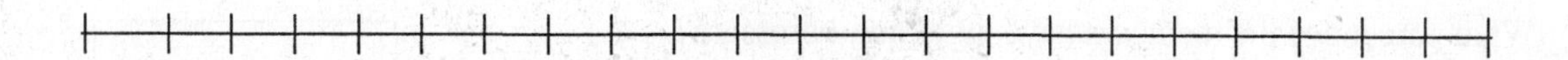

e. Calculate Pearson's coefficient of skewness.

e.

11. The table at the right gives the miles per gallon and the load capacity for 6 SUV's.

a. Develop a scatter diagram for the data.

Miles per Gallons	Load Capacity
19	860
17	970
16	1035
16	1165
14	1180
13	1360

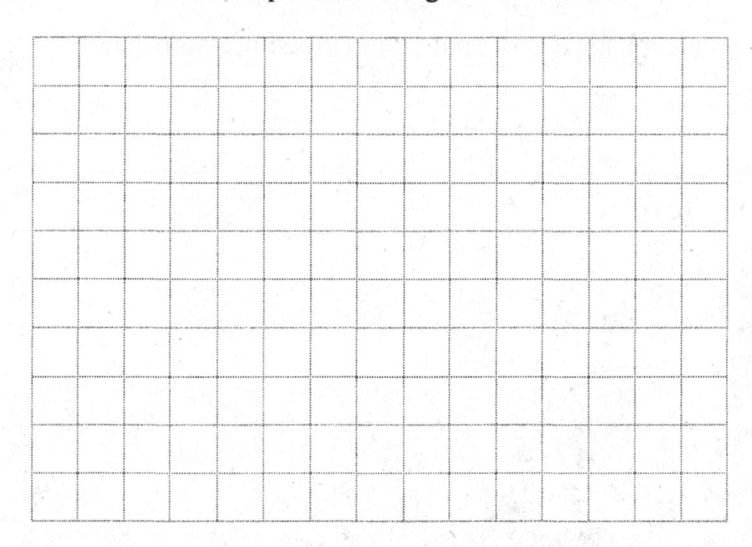

b. How would you characterize the relationship between miles per gallon and load capacity?

12. The Fulton County Farm Bureau held a health screening clinic for its members. The results for the blood pressure screening portion of the clinic are summarized in the table by age group and blood pressure.

Blood	**Age**			
Pressure	**Under 25**	**25 to 50**	**Over 50**	**Total**
Low	20	30	37	87
Medium	44	81	93	218
High	24	45	75	144
Total	88	156	205	449

a. What percent of the members have high blood pressure?

a.

b. What percent of the under age 25 members have low blood pressure?

b.

c. How would you characterize the relationship between medium blood pressure and being over age 50?

____

CHAPTER 5
A SURVEY OF PROBABILITY CONCEPTS

Chapter Goals

When you have completed this chapter, you will be able to:

1. Define probability.
2. Describe the ***classical***, ***empirical***, and ***subjective*** approaches to probability.
3. Understand the terms ***experiment***, ***event***, ***outcome***, ***permutations***, and ***combinations***.
4. Define the terms ***conditional probability*** and ***joint probability***.
5. Calculate probabilities using the ***rules of addition*** and ***rules of multiplication***.
6. Use a ***tree diagram*** to organize and compute probabilities.

Introduction

The emphasis in Chapters 2, 3 and 4 is on ***descriptive statistics***. In those chapters we described methods used to collect, organize, and present data, as well as measures of central location, dispersion, and skewness used to summarize data. A second facet of statistics deals with computing the chance that something will occur in the future. This facet of statistics is called ***inferential statistics***.

An inference is a generalization about a population based on information obtained from a sample. Probability plays a key role in inferential statistics. It is used to measure the reasonableness that a particular sample could have come from a particular population.

What is Probability?

Probability allows us to measure effectively the risks in selecting one alternative over the others. In general, it is a number that describes the chance that something will happen.

> ***Probability***: A value between zero and one, inclusive, describing the relative possibility (chance or likelihood) an event will occur.

Probability is expressed either as a percent or as a decimal. The likelihood that any particular event will happen may assume values between 0 and 1.0. A value close to 0 indicates the event is unlikely to occur, whereas a value close to 1.0 indicates that the event is quite likely to occur.

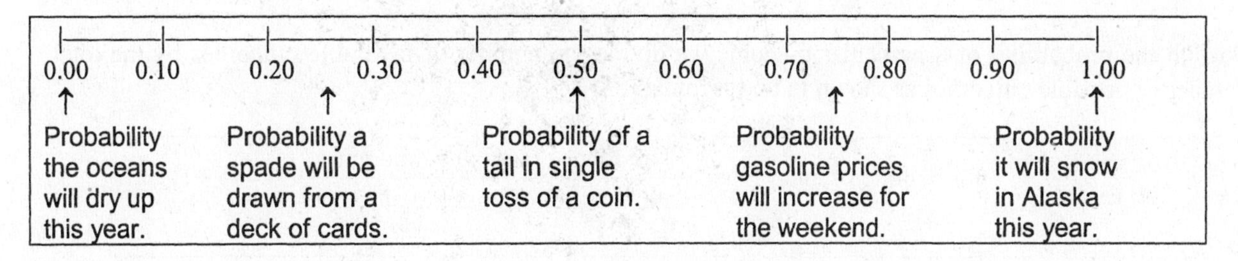

To illustrate, a value of 0.60 might express your degree of belief that tuition will be increased at your college and 0.50 the likelihood that your first marriage will end in divorce.

In our study of probability we will make extensive use of several key words. They are: ***experiment, outcome***, and ***event***.

> ***Experiment***: A process that leads to the occurrence of one and only one of several possible observations.

For example, you roll a die and observe the number of spots that appear face up. The experiment is the act of rolling the die. Your survey company is hired by Ford to poll consumers to determine if they plan to buy a new American-made car this year. You contact a sample of 5,000 consumers. The act of counting the consumers who indicated they would purchase an American-made car is the experiment.

> ***Outcome***: A particular result of an experiment.

One outcome of the die-rolling experiment is the appearance of a 6. In the experiment of counting the number of consumers who plan to buy a new American-made car this year, one possibility is that 2,258 plan to buy a car. Another outcome is that 142 plan to buy one.

> ***Event***: A collection of one or more outcomes of an experiment.

Thus, the ***event*** that the number appearing face up in the die-rolling experiment is an even number is the collection of the ***outcomes*** 2, 4, or 6. Similarly the event that more than half of those surveyed plan to buy a new American made car is the collection of the outcomes 2,501, 2,502, 2,503, and so on all the way up to 5,000.

Approaches To Assigning Probabilities

Two types or classifications of probability are discussed: the objective and subjective viewpoints. Objective probability is subdivided into ***classical probability*** and ***empirical probability***.

Classical Probability

Classical probability is based on the assumption that the outcomes of an experiment are equally likely.

> ***Classical Probability***: A probability based on the assumption that outcomes of an experiment are *equally likely*.

To find the probability of a particular outcome we divide the number of favorable outcomes by the total number of possible outcomes as shown in text formula [5-1].

Probability of an Event

$$\text{Probability of an event} = \frac{\text{Number of favorable outcomes}}{\text{Total number of possible outcomes}} \quad [5-1]$$

For example, you take a multiple-choice examination and have no idea which one of the choices is correct. In desperation you decide to guess the answer to each question. The four choices for each question are the outcomes. They are equally likely, but only one is correct. Thus the probability that you guess a particular answer correctly is 0.25 found by $1 \div 4$.

If only one of several events can occur at one time, we refer to the events as ***mutually exclusive***.

> ***Mutually exclusive:*** The occurrence of one event means that none of the other events can occur at the same time.

An employee selected at random is either a male or female but cannot be both. A computer chip cannot be defective and not defective at the same time.

If an experiment has a set of events that includes every possible outcome, then the set of events is called ***collectively exhaustive.***

> ***Collectively exhaustive:*** At least one of the events must occur when an experiment is conducted.

For example: In a die-tossing experiment every outcome will be either an even number or an odd number. Thus the set is collectively exhaustive. If the set of events is collectively exhaustive and the events are mutually exclusive, the sum of the probabilities equals 1.

Empirical Probability

Another way to define probability is based on ***relative frequencies***. The probability of an event happening is determined by observing what fraction of the time similar events happened in the past.

Probability of an Event Happening

$$\text{Probability of event happening} = \frac{\text{Number of times event occurred in past}}{\text{Total number of observations}}$$

To find a probability using the relative frequency approach we divide the number of times the event has occurred in the past by the total number of observations. Suppose the Civil Aeronautics Board maintained records on the number of times flights arrived late at the Newark International Airport. If 54 flights in a sample of 500 were late, then, according to the relative frequency formula, the probability a particular flight is late is found by:

$$\text{Probability of a late flight} = \frac{\text{number of late flights}}{\text{number of flights}} = \frac{54}{500} = 0.108$$

Based on passed experience, the probability is 0.108 that a flight will be late.

Subjective Probability

If there is little or no past experience or information on which to base a probability, a probability may be determined subjectively. Thus you evaluate the available opinions and other subjective information and then make a decision and arrive at a subjective probability.

> ***Subjective concept of probability***: The likelihood (probability) of a particular event happening that is assigned by an individual based on whatever information is available.

Subjective probability is based on judgment, intuition, or "hunches." The likelihood that the horse, Sir Homer, will win the race at Ferry Downs today is based on the subjective view of the racetrack oddsmaker.

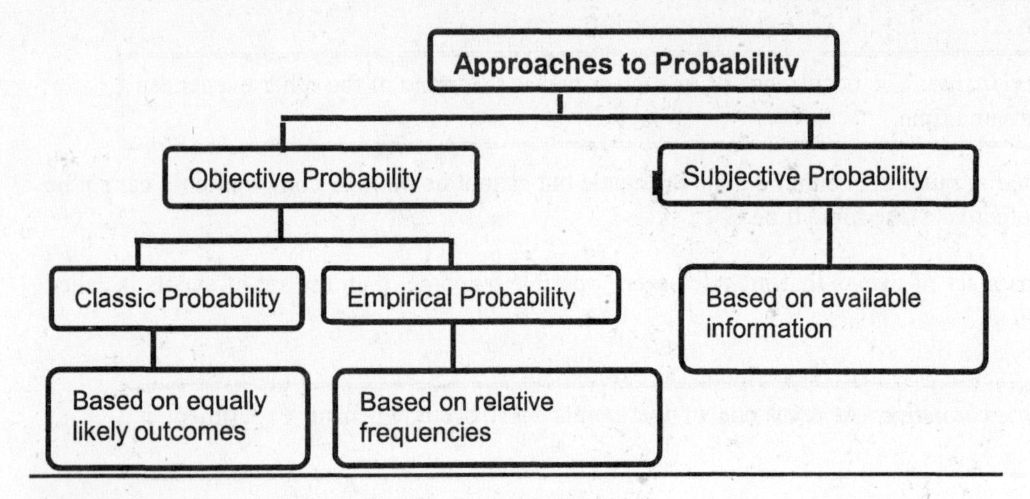

Some Rules for Computing Probabilities

In the study of probability it is often necessary to combine the probabilities of events. This is accomplished through both ***rules of addition*** and ***rules of multiplication***. There are two rules for addition, the ***special rule of addition*** and the ***general rule of addition***.

Special Rule of Addition

To apply the special rule of addition, the events must be *mutually exclusive*. The special rule of addition states

that the probability of the event A ***or*** the event B occurring is equal to the probability of event A plus the probability of event B. The rule is expressed by using text formula [5-2]:

Special Rule of Addition $$P(A\,or\,B)=P(A)+P(B) \qquad [5-2]$$

To apply the special rule of addition the events must be mutually exclusive. This means that when one event occurs none of the other events can occur at the same time.

Venn Diagram

Venn diagrams, developed by English logician J. Venn, are useful for portraying events and their relationship to one another. They are constructed by enclosing a space, usually in a form of a rectangle, which represents the possible events. Two mutually exclusive events such as A and B can then be portrayed—as in the following diagram—by enclosing regions that do not overlap (that is, that have no common area).

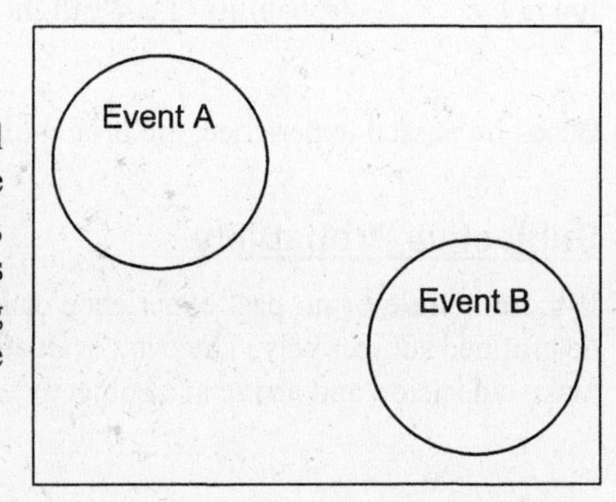

The Complement Rule

The ***complement rule*** is used to determine the probability of an event occurring by subtracting the probability of the event not occurring from one (1).

> ***Complement rule***: A way to determine the probability of an event occurring by subtracting the probability of an event **not** occurring from 1.

This is expressed using text formula [5-3].

Complement Rule $P(A)=1-P(\sim A)$ [5–3]

This formula could be written as:

$$P(A)+P(\sim A)=1$$

In some situations it is more efficient to determine the probability of an event happening by determining the probability of it not happening and subtracting from 1.

General Rule of Addition

When we want to find the probability that two events will both happen, we use the concept known as ***joint probability.***

> ***Joint probability***: A probability that measures the likelihood two or more events will happen concurrently.

What if the events are not mutually exclusive? In that case the general rule of addition is used. The probability is computed using the text formula [5-4].

General Rule of Addition $P(A \text{ or } B)=P(A)+P(B)-P(A \text{ and } B)$ [5–4]

Where:
P(A) is the probability of the event A.
P(B) is the probability of the event B.
P(A and B) is the probability that both events A and B occur.

For example, a study showed 15 percent of the work force to be unemployed, 20 percent of the work force to be minorities, and 5 percent to be both unemployed and minorities. What percent of the work force are either minorities or unemployed?

Note that if *P* (unemployed) and *P* (minority) are totaled, the 5 percent who are both minorities and unemployed are counted in both groups—that is, they are double-counted. They must be subtracted to avoid this double counting. The computation follows.

$$P\text{(unemployed or minority)} = P\text{(unemployed)} + P\text{(minority)} - P\text{(unemployed and minority)}$$
$$= 0.15 + 0.20 - 0.05$$
$$= 0.30$$

These two events are not mutually exclusive and would appear as follows in a Venn diagram:

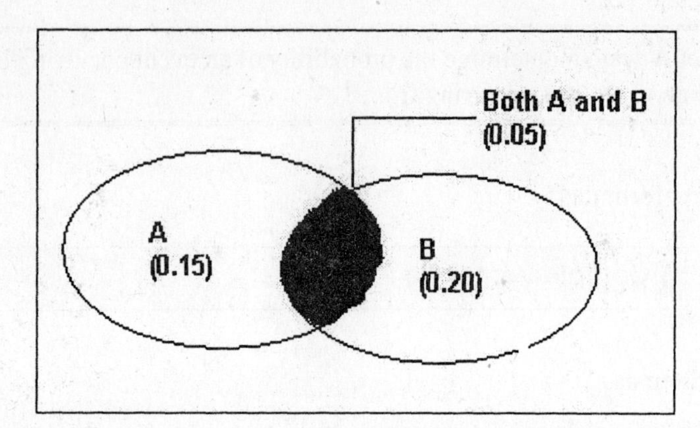

In the above example the likelihood of being both a minority and unemployed is a joint probability.

Using the Addition Rule

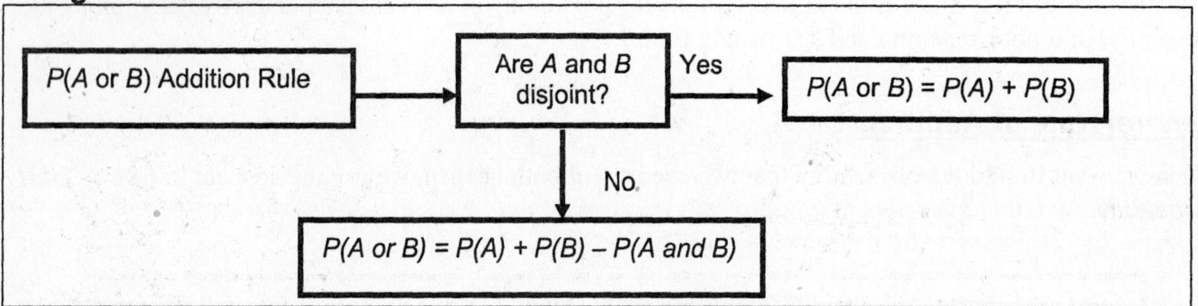

Rules of Multiplication

There were two rules of addition, the general rule and the special rule. We used the general rule when the events were not mutually exclusive and the special rule when the events were mutually exclusive. We have an analogous situation with the rules of multiplication.

Special Rule of Multiplication

We use the general rule of multiplication when the two events are not independent and the ***special rule*** of multiplication when the events **A** and **B** are ***independent***. Two events are independent if the occurrence of one does not affect the probability of the other.

> ***Independence***: The occurrence of one event has no affect on the probability of the occurrence of another event.

A way to look at independence is to assume two events **A** and **B** occur at different times. For example, flipping a coin and getting tails is not affected by rolling a die and getting a two.

The special rule of multiplication is used to combine events where the probability of the second event does not depend on the outcome of the first event.

Another probability concept is called ***conditional probability***.

Conditional probability: The probability of a particular event occurring may be altered by another event that has already occurred.

Probability measures uncertainty, but the degree of uncertainty changes as new information becomes available. Symbolically, it is written $P(B|A)$. The vertical line (|) does not mean divide; it is read "given that" as in the probability of B "given that" A already occurred.

The probability of two independent events, A and B, occurring is found by multiplying the two probabilities. It is written as shown in text formula [5-5].

Special Rule of Multiplication	$P(A \text{ and } B) = P(A) \times P(B)$	[5–5]

As an example, a nuclear power plant has two independent safety systems. The probability the first will not operate properly in an emergency $P(A)$ is 0.01, and the probability the second will not operate $P(B)$ in an emergency is 0.02. What is the probability that in an emergency both of the safety systems will not operate? The probability both will not operate is:

$$\begin{aligned} P(A \text{ and } B) &= P(A) \times P(B) \\ &= 0.01 \times 0.02 \\ &= 0.0002 \end{aligned}$$

The probability 0.0002 is called a joint probability, which is the simultaneous occurrence of two events. It measures the likelihood that two (or more) events will happen together (jointly).

The probability for three independent events, A, B, and C, the special rule of multiplication used to determine the probability of all three events will occur is:

$$P(A \text{ and } B \text{ and } C) = P(A) \times P(B) \times P(C)$$

General Rule of Multiplication

The ***general rule of multiplication*** is used to combine events that are not independent—that is, they are dependent on each other. For two events, the probability of the second event is affected by the outcome of the first event. Under these conditions, the probability of both A and B occurring is given in formula [5-6].

General Rule of Multiplication	$P(A \text{ and } B) = P(A) \times P(B \mid A)$	[5–6]

Where $P(B|A)$ is the probability of B occurring given that A has already occurred. Note that $P(B|A)$ is a conditional probability.

For example, among a group of twelve prisoners, four had been convicted of murder. If two of the twelve are selected for a special rehabilitation program, what is the probability that both of those selected are convicted murderers?

Let A_1 be the first selection (a convicted murderer) and A_2 the second selection (also a convicted murderer). Then $P(A_1) = 4/12$. After the first selection, there are 11 prisoners, 3 of whom are convicted of murder, hence $P(A_2|A_1) = 3/11$. The probability of both A_1 and A_2 happening is:

$$\begin{aligned} P(A_1 \text{ and } A_2) &= P(A_1) \times P(A_2|A_1) \\ &= \frac{4}{12} \times \frac{3}{11} \\ &= 0.0909 \end{aligned}$$

Contingency Tables

Recall from Chapter 4, that if we tally the results of a survey into a two-way table, the results of this tally can be used to determine various possibilities. This two-way table is referred to as a ***contingency table.***

> ***Contingency table:*** A table used to classify sample observations according to two or more identifiable characteristics.

A contingency table is a cross tabulation that simultaneously summarizes two variables of interest and their relationship.

At the right is an example.

	Gender		
Bought Pepsi	**Boys**	**Girls**	**Total**
0	5	10	15
1	15	25	40
2 or more	80	65	145
Total	100	100	200

A survey of 200 school children classified each as to gender and the number of times Pepsi-Cola was purchased each month at school. Each respondent is classified according to two criteria-the number of times Pepsi was purchased and gender.

Using the Multiplication Rule

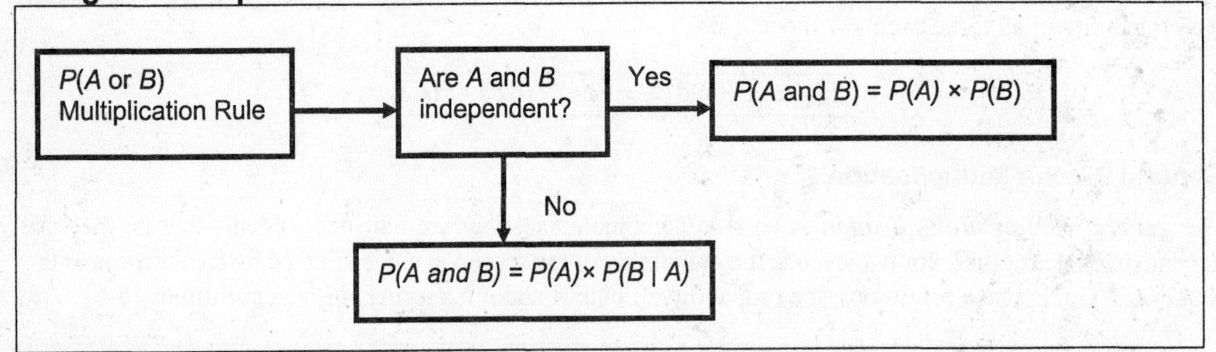

Tree Diagrams

The ***tree diagram*** is a graph that is helpful in organizing the calculations that involve several stages. Each segment in the tree is one stage of the problem. The branches are weighted probabilities.

Principles of Counting

If the number of possible outcomes in an experiment is small, it is relatively easy to count the possible outcomes. However, sometimes the number of possible outcomes is large, and listing all the possibilities would be time consuming, tedious, and error prone. Three formulas are very useful for determining the number of possible outcomes in an experiment. They are: the ***multiplication formula***, the ***permutation formula***, and the ***combination formula***.

The Multiplication Formula

The general rule of multiplication is used to combine events that are dependent on each other.

Multiplication formula: If there are m ways of doing one thing, and n ways of doing another thing, there are $m \times n$ ways of doing both.

In terms of a formula:

Multiplication Formula	$\text{Total number of arrangements} = (m)(n)$	[5–7]

The Permutation Formula

The permutation is an arrangement of objects or things wherein order is important. That is, each time the objects or things are placed in a different order, a new permutation results.

Permutation: Any arrangements of r objects selected from a single group of n possible objects.

The formula for the number of permutations is:

Permutation Formula	$_nP_r = \frac{n!}{(n-r)!}$	[5–8]

Where:

P is the number of permutations, or ways the objects can be arranged.
n is the total number of objects.
r is the number of objects selected.

Note: the $n!$ is a notation called "n factorial."

For example, 4! means 4 times 3 times 2 times 1 $= 4 \times 3 \times 2 \times 1 = 24$

This concept is shown in Example 7.

The Combination Formula

One particular arrangement of the objects without regard to order is called a ***combination***.

Combination: The number of ways to choose r objects from a group of n possible objects without regard to order.

The formula for the number of combinations is:

Combination Formula	$_nC_r = \frac{n!}{r!(n-r)!}$	[5–9]

Where:
C is the number of different combinations.
n is the total number of objects.
r is the number of objects to be used at one time.

This concept is shown in Example 8.

Glossary

Classical Probability: A probability based on the assumption that the outcomes for an experiment are equally likely.

Collectively exhaustive: At least one of the events must occur when an experiment is conducted.

Combination: The number of ways to choose *r* objects from a group of *n* possible objects without regard to order.

Complement rule: A way to determine the probability of an event occurring by subtracting the probability of an event **not** occurring from 1.

Conditional probability: The probability of a particular event occurring, given that another event occurred.

Contingency table: A table used to classify sample observations according to two or more identifiable characteristics.

Event: A collection of one or more outcomes of an experiment.

Experiment: A process that leads to the occurrence of one and only one of several possible observations.

Independence: The occurrence of one event has no affect on the probability of the occurrence of another event.

Joint probability: A probability that measures the likelihood two or more events will happen concurrently.

Multiplication formula: If there are *m* ways of doing one thing, and *n* ways of doing another thing, there are $m \times n$ ways of doing both.

Mutually exclusive: The occurrence of one event means that none of the other events can occur at the same time.

Outcome: A particular result of an experiment.

Permutation: Any arrangements of *r* objects selected from a single group of *n* possible objects.

Probability: A value between zero and one, inclusive, describing the relative possibility (chance or likelihood) an event will occur.

Subjective concept of probability: The likelihood (probability) of a particular event happening that is assigned by an individual based on whatever information is available.

Chapter Examples

Example 1 – Finding the Probability of an Event

Dunn Pontiac has compiled the following sales data regarding the number of cars sold over the past 60 selling days. Answer the following questions for the sales data shown.

Dunn Pontiac Sales Data	
Number of Cars Sold	Number of Days
0	5
1	5
2	10
3	20
4	15
5 or more	5
Total	60

a. What is the probability that two cars are sold during a particular day?

b. What is the probability of selling 3 or more cars during a particular day?

c. What is the probability of selling at least one car during a particular day?

Solution 1 – Finding the Probability of an Event

This problem is an example of the relative frequency type of probability, because the probability of an event happening is based on the number of times the particular event happened in the past relative to the total number of observations.

a. The probability that exactly two cars are sold is:

$$P(2\,\text{cars}) = \frac{\text{Number of days two cars were sold}}{\text{Total number of days}} = \frac{10}{60} = 0.17$$

b. The probability of selling three or more cars is found by using the special rule of addition given in formula [5-2]. Let X represent the number of cars sold. ($\geq$ is read "greater than or equal to." The notation $>$ would be just greater than.) Then:

$$P(X \geq 3) = P(3) + P(4) + P(5\,\text{or more}) = \frac{20}{60} + \frac{15}{60} + \frac{5}{60} = \frac{40}{60} = 0.67$$

Interpreting, three cars or more are sold 67 percent of the days.

c. The probability of selling at least one car is determined by adding the probabilities of selling one, two, three, four, and five or more cars. Again let X be the number of cars sold, then

$$P(X \geq 1) = P(1) + P(2) + P(3) + P(4) + P(5\,\text{or more})$$
$$= \frac{5}{60} + \frac{10}{60} + \frac{20}{60} + \frac{15}{60} + \frac{5}{60} = \frac{55}{60} = 0.9166 = 0.92$$

The same result can also be found by using the complement rule. Obtain the probability of the occurrence of a particular event by computing the probability it did not occur and then subtracting that value from 1.0. In this example, the probability of not selling any cars is $\frac{5}{60} = 0.083 = .08$, then $(1 - 0.08) = 0.92$.

Self-Review 5.1

Check your answers against those in the ANSWER section.

A study was made to investigate the number of times adult males over 30 visit a physician each year. The results for a sample of 300 were:

Number of Visits	Number of Adult Males
0	30
1	60
2	90
3 or more	120
Total	300

a. What is the probability of selecting someone who visits a physician twice a year?

b. What is the probability of selecting someone who visits a physician?

Example 2 – Finding Probability Using the Rule of Addition

A local community has two newspapers. The ***Morning Times*** is read by 45 percent of the households. The ***USA Today*** is read by 60 percent of the households. Twenty percent of the households read both papers. What is the probability that a particular household in the city reads at least one paper?

Solution 2 – Finding Probability Using the Rule of Addition

If we combine the probabilities (0.45, 0.60, and 0.20), they exceed 1.00. The group that reads both papers, of course, is being counted twice and must be subtracted to arrive at the answer. Let T represent the ***Morning Times***, and U represent ***USA Today***, and use the general rule of addition, formula [5-4]:

$$
\begin{aligned}
P(T \text{ or } U) &= P(T) + P(U) - P(T \text{ and } U) \\
&= 0.45 + 0.60 - 0.20 \\
&= 0.85
\end{aligned}
$$

Thus, 85 percent of the households in the community read at least one paper.

Self-Review 5.2

Check your answers against those in the ANSWER section.

The proportion of students at Pemberville University who own an automobile is 0.60. The proportion that lives in a dormitory is 0.20. The proportion of students that both own an automobile and live in a dormitory is 0.12. What proportion of the students either owns an auto or lives in a dorm?

Example 3 – Finding Probability Using the Special Rule of Multiplication

The probability that a bomber hits a target on a bombing mission is 0.70. Three bombers are sent to bomb a particular target. What is the probability that they all hit the target? What is the probability that at least one hits the target?

Solution 3 – Finding Probability Using the Special Rule of Multiplication

These events are independent since the probability that one bomber hits the target does not depend on whether the other hits it. The special rule of multiplication, formula [5-5], is used to find the joint probability. B_1 represents the first bomber, B_2 the second bomber, and B_3 the third bomber.

$$P(\text{all 3 hit target}) = P(B_1) \times P(B_2) \times P(B_3)$$
$$= (0.70)(0.70)(0.70)$$
$$= 0.343$$

Hence the probability that all three complete the mission is 0.343.

The probability that at least one bomber hits the target is found by combining the complement rule and the multiplication rule. To explain: The probability of a miss with the first bomber is 0.30, found by $P(M_1) = 1 - 0.70$. The probability for M_2 and M_3 is also 0.30. The multiplication rule is used to obtain the probability that all three miss. Let X be the number of hits.

$$P(X > 0) = 1 - P(0)$$
$$= 1 - [P(M_1)][P(M_2)][P(M_3)]$$
$$= 1 - [(0.30)(0.30)(0.30)]$$
$$= 1 - 0.027 = 0.973$$

Thus, the likelihood that at least one of the bombs hit the target is 0.973.

Self-Review 5.3

Check your answers against those in the ANSWER section.

A side effect of a certain anesthetic used in surgery is the hiccups, which occurs in about 10 percent of the cases. If three patients are scheduled for surgery today and are to be administered this anesthetic, compute the probability that:

a. three get hiccups **b.** none get hiccups **c.** at least one gets hiccups

Example 4 – Finding a Probability

Yesterday, the Bunte Auto Repair Shop received a shipment of four carburetors. One is known to be defective. If two are selected at random and tested:

a. What is the probability that neither one is defective?

b. What is the probability that the defective carburetor is located by testing two carburetors?

Solution 4 – Finding a Probability

a. The selections of the two carburetors are not independent events because the selection of the first affects the second outcome. Use formula [5-6]. Let G_1 represent the first "good" carburetor and G_2 the second "good" one.

$P(G_1 \text{ and } G_2) = P(G_1) \times P(G_2|G_1)$

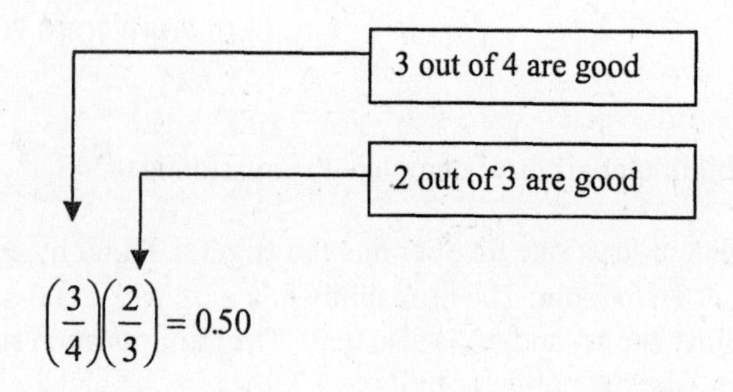

$$\left(\frac{3}{4}\right)\left(\frac{2}{3}\right) = 0.50$$

Hence, the probability that neither of the two selected carburetors is defective is 0.50.

b. The probability that the defective carburetor is found requires the general rule of multiplication and the general rule of addition.

In this case the defect may be detected either in the first test or in the second one. The general rule of multiplication is used. Let D_1 represent a defect on the first test and D_2 on the second test. The probability is:

$$P(\text{find the defect}) = P(G_1) \times P(D_2 \mid G_1) + P(D_1) \times P(G_2 \mid D_1)$$
$$= \left(\frac{3}{4}\right)\left(\frac{1}{3}\right) + \left(\frac{1}{4}\right)\left(\frac{3}{3}\right) = \frac{1}{4} + \frac{1}{4} = \frac{1}{2} = 0.50$$

To explain further, the probability that the first carburetor tested is good is $P(D_1) = ¾$. If the first one selected is good, then to meet the requirements of the problem the second one sampled must be defective. This conditional probability is $P(D_2|G_1) = 1/3$. The joint probability of these two events is 3/12 or ¼. The defective part could be found on the first test $P(D_1)$. Since there is one defect among the four carburetors, the probability that it will be found on the first test is ¼. If the defect is found on the first test, then the three remaining parts are good. Hence the conditional probability of selecting a good carburetor on the second trial is 1.0 $P(G_2|D_1)$. The joint probability of a defective part being followed by a good part is ¼, found by $P(D_1)$ x $P(G_2|D_1) = (1/4)(3/3) = ¼$. The sum of these two outcomes is 0.50.

Self-Review 5.4

Check your answers against those in the ANSWER section.

Ten students are being interviewed for a class office. Six of them are female and four are male. Their names are all placed in a box and two students are selected for the interviews.

a. What is the probability that both of those selected are female?

b. What is the probability that at least one is male?

Example 5 – Finding Conditional Probability

A large department store is analyzing the per-customer amount of purchase and the method of payment. For a sample of 140 customers, the following contingency table or cross-classified table presents the findings.

Payment Method	Amount of Purchase			Total
	B_1: Less than $20	B_2:$20 up to $50	B_3: $50 or more	
A_1: Cash	15	10	5	30
A_2: Check	10	30	20	60
A_3:Charge	10	20	20	50
Total	35	60	45	140

a. What is the probability of selecting someone who paid by cash or made a purchase of less than $20?

b. What is the probability of selecting someone who paid by check and made a purchase of more than $50?

Solution 5 – Finding Conditional Probability

a. If we combine the events "Less than $20" ($B_1$) and "Cash payment" ($A_1$), then those who paid cash for a purchase of less than $20 are counted twice. That is, these two events are not mutually exclusive. Therefore, the general rule of addition formula [5-4] is used.

$$P(A_1 \text{ or } B_1) = P(A_1) + P(B_1) - P(A_1 \text{ and } B_1)$$
$$= \frac{30}{140} + \frac{35}{140} - \frac{15}{140} = \frac{50}{140} = 0.36$$

The probability of selecting a customer who made a cash payment or purchased an item for less than $20 is 0.36.

b. Conditional probability is used to find the probability of selecting someone who paid by check (A_2) and who made a purchase of over $50 ($B_3$)

There are two qualifications: "paid by check" and "made a purchase of over $50." Referring to the table, 20 out of 140 customers meet both qualifications, therefore, $20 \div 140 = 0.14$.

This probability could also be computed in a three-step process.

1. The probability of selecting those who paid by check (A_2) is $60 \div 140 = 0.43$.
2. Of the 60 persons who paid by check, 20 made a purchase of over $50. Therefore $P(B_3|A_2) = 20 \div 60 = 0.33$
3. These two events are then combined using the general rule of multiplication, formula [5-6].

$$P(A_2 \text{ and } B_3) = P(A_2) \times P(B_3|A_2)$$
$$= (0.43)(0.33) = 0.14$$

Self-Review 5.5

Check your answers against those in the ANSWER section.

Five hundred adults over 50 years of age were classified according to whether they smoked or not, and if they smoked, were they a moderate or heavy smoker. Also, each one was asked whether he or she had ever had a heart attack. The results are given.

	Heart Attack		
	Yes	No	Total
Do not smoke	30	220	250
Moderate smoker	60	65	125
Heavy smoker	90	35	125
Totals	180	320	500

a. What is the probability of selecting a person who either has had a heart attack, or who is a heavy smoker?

b. What is the probability of selecting a heavy smoker who did not have a heart attack?

Example 6 – Determining Probability Using the Multiplication Formula

A deli bar offers a special sandwich for which there is a choice of five different cheeses, four different meat selections, and three different rolls. How many different sandwich combinations are possible?

Solution 6 – Determining Probability Using the Multiplication Formula

Using the multiplication formula [5-7], there are five cheeses (*c*), four meats (*m*), and three rolls (*r*). The total number of possible sandwiches is 60 found by:

$$cmr = (5)(4)(3) = 60$$

Self-Review 5.6

Check your answers against those in the ANSWER section.

The Swansons are planning to fly to Hawaii from Toronto with a stopover in Los Angeles. There are five flights they can take between Toronto and Los Angeles and ten flights between Los Angeles and Hawaii. How many different flights are possible between Toronto and Hawaii?

Example 7 – Determining a Permutation

Three scholarships are available for needy students. Their values are: $2,000, $2,400, and $3,000. Twelve students have applied and no student may receive more than one scholarship. Assuming all twelve students are in need of funds, how many different ways could the scholarships be awarded?

Solution 7 – Determining a Permutation

This is an example of a permutation because a different assignment of the scholarships means another arrangement. Jones could be awarded the $2,000 scholarship, Sinski the $2,400 scholarship, and Peters the $3,000 scholarship. Or, Sinski could be awarded the $2,000, Seiple the $2,400 one, and Orts the $3,000 scholarship, and so on. Using formula [5-8]:

$${}_nP_r = \frac{n!}{(n-r)!} = \frac{12!}{(12-3)!} = \frac{12\times 11\times 10\times 9!}{9!} = 12\times 11\times 10 = 1{,}320$$

Where:
n is the total number of applicants
r is number of scholarships.
Note that the 9! in the numerator and the 9! in the denominator cancel each other.

Self-Review 5.7
Check your answers against those in the ANSWER section.

Swanton Welding is setting up the shop for a new production run. Nine welding machines are available but only three spaces are available in the production area of the shop. In how many different ways can the nine welding machines be arranged in the three available spaces?

Example 8 – Determining a Combination

The basketball coach of Dalton University is quite concerned about their 40 straight losses. The frustrated coach decided to select the starting lineup for the DU-UCLA game by drawing five names from the 12 available players at random. (Assume that a player can play any position.) How many different starting lineups are possible?

Solution 8 – Determining a Combination

This is an example of a combination because the order in which the players are selected is not important. Jocko, Camden, Urfer, Smith, and Marchal are the same starting lineup as Smith, Camden, Marchal, Jocko, and Urfer, and so on. Using formula [5-9]:

$${}_nC_r = \frac{n!}{r!(n-r)!} = \frac{12!}{5!(12-5)!} = \frac{12 \times 11 \times 10 \times 9 \times 8 \times 7!}{5 \times 4 \times 3 \times 2 \times 1 \times 7!} = \frac{95{,}040}{120} = 792$$

Where:
n is the total number of available players.
r is the number in the starting lineup.

Self-Review 5.8
Check your answers against those in the ANSWER section.

A major corporation has branch offices in eight major cities in the United States and Canada. The company president wants to visit five of these offices. How many different trip combinations are possible?

CHAPTER 5 ASSIGNMENT

A SURVEY OF PROBABILITY CONCEPTS

Name ______________________________ Section _________ Score _______

Part I Select the correct answer and write the appropriate letter in the space provided.

______ **1.** Which of the following statements regarding probability is always correct?
- **a.** A probability can range from 0 to 1.
- **b.** A probability close to 0 means the event is not likely to happen.
- **c.** A probability close to 1 means the event is likely to happen.
- **d.** all of the above are correct.

______ **2.** According to the classical definition of probability
- **a.** All the events are equally likely.
- **b.** The probability is based on hunches.
- **c.** Divide the number of successes by the number of failures.
- **d.** One outcome is exactly twice the other.

______ **3.** The observation of some activity or the act of taking some measurement is called
- **a.** an outcome.
- **b.** an experiment.
- **c.** a probability.
- **d.** an event.

______ **4.** The particular result of an experiment is called
- **a.** an experiment.
- **b.** an event.
- **c.** a probability.
- **d.** an outcome.

______ **5.** An event is the collection of one or more
- **a.** outcomes.
- **b.** combinations.
- **c.** probabilities.
- **d.** experiments.

______ **6.** If A and B are mutually exclusive events then $P(A \text{ or } B)$ equals
- **a.** $P(A) + P(B) - P(A \text{ and } B)$
- **b.** $P(A) \times P(B)$
- **c.** $P(A) + P(B)$
- **d.** $P(A|B) + P(B|A)$

______ **7.** If A and B are independent events, then $P(A \text{ and } B))$ equals
- **a.** $P(A) + P(B|A)$.
- **b.** $P(A) \times P(B)$.
- **c.** $P(A) + P(B)$.
- **d.** $P(A|B) + P(B|A)$.

______ **8.** Which formula represents the probability of the complement of event A?
- **a.** $1 + P(A)$
- **b.** $1 - P(A)$
- **c.** $P(A)$
- **d.** $P(A) - 1$

______ **9.** The simultaneous occurrence of two events is called
- **a.** prior probability
- **b.** subjective probability
- **c.** conditional probability
- **d.** joint probability

______ **10.** If the probability of an event is 0.3, that means

a. the event has a 70% chance of not occurring.

b. the complement of the event has a 30% chance of occurring.

c. the event has a 30% chance of not occurring.

d. the complement of the event has a 70% chance of not occurring.

Part II Answer each question below. Be sure to show all of your work.

11. A recent study of young executives showed that 30 percent run, 20 percent bike and 12 percent do both. What is the percent of young executives who run or bike?

11.

12. A survey of publishing jobs indicates that 92 percent are completed on time. Assume that three jobs are selected for study.

a. What is the probability they are all completed on time?

a.

b. What is the probability that at least one was not completed on time?

b.

13. Today's local newspaper lists 20 stocks "of local interest." Of these stocks, ten increased, five decreased and five remained unchanged yesterday. If we decide to buy two of the stocks, what is the likelihood that both increased yesterday?

13.

14. Six employees of a marketing firm had effectiveness ratings as follows: 0.72, 0.46, 0.59, 0.64, 0.81 and 0.76. Find the probability of selecting an employee with the indicated effectiveness rating.

a. Greater than 0.75.

a.

b. Less than 0.75 but greater than 0.5.

b.

c. Greater than the mean.

c.

d. Not less than 0.9.

d.

15. A freight train is going to carry four tankers, six coal cars, and ten lumber cars. How many different arrangements of cars are possible?

15.

16. The United Way Campaign of Greater Toledo had fifteen applications for funding this year. If eight of these applications can be funded, how many different lists of successful applications are there?

16.

17. A market analyst is hired to provide information on the type of customers who shop at a particular store. A random survey is taken of 100 shoppers at this store. Of these 100, 73 are women. The shoppers are grouped in three age categories, under 30, 30 up to 50 and 50 and over. The data is summarized in the table.

	Women	Men	Total
Under 30	30	8	38
30 to 50	25	14	39
50 & over	18	5	23
Totals	73	27	100

Let W be the event that a randomly selected shopper is a woman.
Let A be the event that a randomly selected shopper is under 30.

a. Find the probability of $\sim W$.

a.

b. Find the probability of A.

b.

c. Find the probability of A and W.

c.

d. Find the probability of A or $\sim W$.

d.

e. Find the probability of A given W.

e.

CHAPTER 6
DISCRETE PROBABILITY DISTRIBUTIONS

Chapter Goals

After completing this chapter, you will be able to:

1. Define the terms ***probability distribution*** and ***random variable***.
2. Distinguish between ***discrete*** and ***continuous probability distributions***.
3. Calculate the mean, variance, and standard deviation of a ***discrete probability distribution***.
4. Describe the characteristics of and compute probabilities using the ***binomial probability distribution***.
5. Describe the characteristics of and compute probabilities using the ***Poisson probability distribution***.

Introduction

In the previous chapter we discussed the basic concepts of probability and described how the rules of addition and multiplication were used to compute probabilities. In this chapter we expand the study of probability to include the concept of a ***probability distribution***.

What is a Probability Distribution?

A probability distribution shows the possible outcomes of an experiment and the probability of each of these outcomes.

> ***Probability distribution***: A listing of all the outcomes of an experiment and the probability associated with each outcome.

How can we generate a probability distribution? As an example, the possible outcomes on the roll of a single die are shown at right.

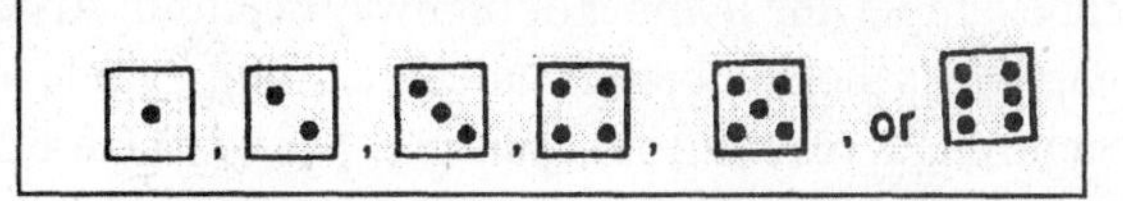

Each face should appear on about one-sixth of the rolls. The table shows the possible outcomes and corresponding probabilities for this experiment. It is a discrete distribution because only certain outcomes are possible and the distribution is a result of counting the various outcomes.

Number of Spots on Die	Probability Fraction		Decimal
1	1/6	=	0.1667
2	1/6	=	0.1667
3	1/6	=	0.1667
4	1/6	=	0.1667
5	1/6	=	0.1667
6	1/6	=	0.1667
Total	6/6	=	1.0002

There are several important features of the discrete probability distribution:

1. The listing is exhaustive; that is, all the possible outcomes are included.
2. The total (sum) of all possible outcomes is 1.0. (In the example, total is greater than 1 because of rounding.)
3. The probability of any particular outcome is between 0 and 1 inclusive. (See the table above.)

4. The outcomes are mutually exclusive, meaning, for example, a 6 spot and a 2 spot cannot appear at the same time on the roll of one die.

This discrete probability distribution, presented above as a table, may also be portrayed in graphic form as shown on the right.

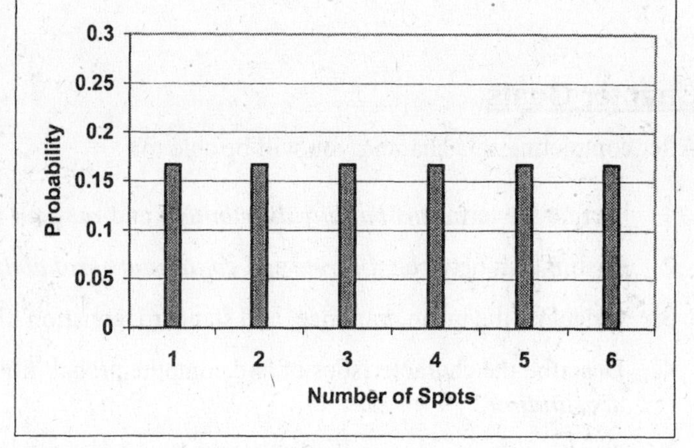

By convention the probability is shown on the Y-axis (the vertical axis) and the outcomes on the X-axis (the horizontal axis). This probability distribution is often referred to as a uniform distribution.

A probability distribution can also be expressed in equation form.

For example:

$P(x) = 1/6$, where x can assume the values 1, 2, 3, 4, 5, or 6.

Random Variable

In any experiment of chance, the outcomes occur randomly. A ***random variable*** is a value determined by the outcome of an experiment.

> ***Random Variable*:** A quantity resulting from an experiment that, by chance, can assume different values.

A random variable may have two forms: discrete or continuous. A ***discrete random variable*** may assume only distinct values and is usually the result of counting.

> ***Discrete random variable*:** A variable that can assume only certain clearly separated values.

For example, the number of highway deaths in Arkansas on Memorial Day weekend may be 1, 2, 3,... Another example is the number of students earning a grade of B in your statistics class. In both instances the number of occurrences results from counting. Note that there can be 12 deaths or 15 B's but there cannot be 12.63 deaths or 15.27 B grades.

If we measure something, such as the diameter of a tree, the length of a field, or the time it takes to run the Boston Marathon, the variable is called a ***continuous random variable***.

> ***Continuous random variable*:** A variable that can assume one of an infinitely large number of values within certain limitations.

In brief, if the problem involves counting something, the resulting distribution is usually a discrete probability distribution. If the distribution is the result of a measurement, then it is usually a continuous probability distribution.

What is the difference between a random variable and a probability distribution? A probability distribution lists all the possible outcomes as well as their corresponding probabilities. A random variable lists only the outcomes. We will examine the continuous random variable and the continuous probability distribution in the next chapter.

The Mean, Variance, and Standard Deviation of a Probability Distribution

In Chapter 3 we computed the mean and variance of a frequency distribution. The mean is a measure of location and the variance is a measure of the spread of the data. In a similar fashion the ***mean*** (μ) and the ***variance*** (σ^2) summarize a probability distribution.

The Mean

The ***mean*** μ, or expected value $E(x)$, is used to represent the central location of a probability distribution. It is also the long-run average value of the random variable. It is computed by the following formula:

Mean of a Probability Distribution	$\mu = \Sigma\left[xP(x)\right]$	[6−1]

This formula directs you to multiply each outcome (x) by its probability $P(x)$; and then add the products.

Variance and Standard Deviation

While the mean is a typical value used to summarize a discrete probability distribution, it does not tell us anything about the spread in the distribution. The ***variance*** tells us about the spread or variation in the data. The variance is computed using the following formula:

Variance of a Probability Distribution	$\sigma^2 = \Sigma\left[(x-\mu)^2 P(x)\right]$	[6−2]

The steps in computing the variance using formula [6-2] are:

1. Subtract the mean (μ) from each outcome (x) and square these differences.
2. Multiply each squared difference by its probability $P(x)$
3. Sum these products to arrive at the variance.

The ***standard deviation*** (σ) of a discrete probability distribution is found by calculating the positive square root of σ^2, thus $\sigma = \sqrt{\sigma^2}$.

Binomial Probability Distribution

One of the most widely used discrete probability distributions is the ***binomial probability distribution***. It has the following characteristics:

1. An outcome on each trial of an experiment is classified into one of two mutually exclusive categories — a success or a failure.
2. The random variable is the number of successes in a fixed number of trials.
3. The probability of success and failure stay the same for each trial.

4. The trials are independent, meaning that the outcome of one trial does not affect the outcome of any other trial.

Illustrations of each characteristic are:

1. Each outcome is classified into one of two mutually exclusive categories. An outcome is classified as either a "success" or a "failure." For example, 40 percent of the students at a particular university are enrolled in the College of Business. For a selected student there are only two possible outcomes–the student is enrolled in the College of Business (designated a success) or he/she is not enrolled in the College of Business (designated a failure).
2. The binomial distribution is the result of counting the number of successes in a fixed sample size. If we select 5 students, 0, 1, 2, 3, 4, or 5 could be enrolled in the College of Business. This rules out the possibility of 3.45 of the students being enrolled in the College of Business. That is, there cannot be fractional counts.
3. The probability of a success remains the same from trial to trial. In the example regarding the College of Business, the probability of a success remains at 40 percent for all five students selected.
4. Each sampled item is independent. This means that if the first student selected is enrolled in the College of Business, it has no effect on whether the second or the fourth one selected will be in the College of Business.

How a Binomial Probability Distribution is Computed

To construct a binomial probability distribution we need to know:

(1) The number of trials, designated n.
(2) The probability of success (π) on each trial.

The binomial probability distribution is constructed using formula [6-3]:

Binomial Probability Distribution $$P(x) = {}_nC_x(\pi)^x(1-\pi)^{n-x} \quad [6-3]$$

Where:
${}_nC_x$ denotes a combination of n items selected x at a time
n is the number of trials
x is the random variable defined as the number of observed successes
π is the probability of success on each trial (Do not confuse it with the mathematical constant 3.1416.)

The mean (μ) and variance (σ^2) of a binomial distribution can be computed by these formulas.

Mean of a Binomial Distribution $$\mu = n\pi \quad [6-4]$$

Variance of a Binomial Distribution $$\sigma^2 = n\pi(1-\pi) \quad [6-5]$$

Poisson Probability Distribution

Another discrete probability distribution is the ***Poisson probability distribution***.

> ***Poisson probability distribution***: Has the same four characteristics as the binomial, but in addition the probability of success (π) is small, and n, the number of trials, is relatively large.

Charactertistics of the Poisson Probability Distribution:

1. The random variable is the number of times some event occurs during a defined interval.
2. The probability of the event is proportional to the size of the interval.
3. The intervals do not overlap and are independent.

The formula for computing the probability of a success is:

Poisson Distribution $$P(x) = \frac{\mu^x e^{-\mu}}{x!} \quad [6-6]$$

Where:

$P(x)$ is the probability for a specified value of x.
x is the number of occurrences (successes).
μ is the arithmetic mean number of occurrences (successes) in a particular interval.
e is the mathematical constant 2.71828. (base of the Napierian logarithm system)

Note that the mean number of successes, μ, can be determined in binomial situations by $n\pi$, where n is the total number of trials and π is the probability of success.

Mean of a Poisson Distribution $$\mu = n\pi \quad [6-7]$$

As an example where the Poisson distribution is applicable, suppose electric utility statements are based on the actual reading of the electric meter. In 1 out of 100 cases the meter is incorrectly read ($\pi = 0.01$). Suppose the number of errors that appear in the processing of 500 customer statements approximates the Poisson distribution ($n = 500$). In this case the mean number of incorrect statements is found by:

$$\mu = n\pi = 500\,(0.01) = 5$$

Using formula [5-6], finding the probability of exactly two errors appearing in 500 customer statements is rather tedious. Instead we merely refer to the Poisson distribution in Appendix C. Locate by μ = (5.0) at the top of a set of columns. Then find the x of 2 in the left column and read across to the column headed by 5.0. The probability of exactly 2 statement errors is 0.0842.

GLOSSARY

Continuous random variable: A variable that can assume one of an infinitely large number of values within certain limitations.

Discrete random variable: A random variable that can assume only certain clearly separated values.

Poisson probability distribution: Has the same four characteristics as the binomial, but in addition the probability of success (π) is small, and n, the number of trials, is relatively large.

Probability distribution: A listing of all the outcomes of an experiment and the probability associated with each outcome.

Random Variable: A quantity resulting from an experiment that, by chance, can assume different values.

CHAPTER EXAMPLES

Example 1 – Compute Arithmetic Mean and Variance of a Probability Distribution

Number of Breakdowns Per Week	Weeks	Probability		Probability
0	20	20/60	=	0.333
1	20	20/60	=	0.333
2	10	10/60	=	0.167
3	10	10/60	=	0.167
Total	60			1.000

Bill Russe, production manager at Ross Manufacturing, maintains detailed records on the number of times each machine breaks down and requires service during the week. Bill's records show that the Puret grinder has required repair service according to the following distribution. Compute the arithmetic mean and the variance of the number of breakdowns per week.

Solution 1 – Compute Arithmetic Mean and Variance of a Probability Distribution

Number of Breakdowns Per Week	Probability	
x	$P(x)$	$xP(x)$
0	0.333	0.000
1	0.333	0.333
2	0.167	0.334
3	0.167	0.501
	Total	1.168

The arithmetic mean, or expected number of breakdowns per week for the probability distribution, is computed using formula [6-1].

$$\mu = \Sigma\left[xP(x)\right] \qquad [6-1]$$

The arithmetic mean number of times the Puret machine breaks down per week is 1.168. The variance of the number of breakdowns is computed using formula [6-2].

$$\sigma^2 = \Sigma\left[(x-\mu)^2 P(x)\right] \qquad [6-2]$$

Number of Breakdowns Per Week	Probability			
x	$P(x)$	$(x-\mu)$	$(x-\mu)^2$	$(x-\mu)^2 P(x)$
0	0.333	0 – 1.168 = –1.168	1.364	(1.364)(0.333) = 0.454212
1	0.333	1 – 1.168 = –0.168	0.028	(0.028)(0.333) = 0.009324
2	0.167	2 – 1.168 = +0.832	0.692	(0.692)(0.167) = 0.115564
3	0.167	3 – 1.168 = +1.832	3.356	(3.356)(0.167) = 0.560452
				Total 1.139552

The variance of the number of breakdowns per week is about 1.140. The standard deviation of the number of breakdowns per week is 1.07, found by $\sqrt{1.139552} = 1.0674 = 1.07$.

Self-Review 6.1

Check your answers against those in the ANSWER section.

The safety engineer at Manellis Electronics reported the following probability distribution for the number of on-the-job accidents during a one-month period.

a. Compute the mean **b.** Compute the variance.

Number of Accidents	Probability
0	0.60
1	0.30
2	0.10

Example 2 – Using the Rules for a Binomial Probability Distribution

An insurance representative has appointments with four prospective clients tomorrow. From past experience she knows that the probability of making a sale on any appointment is 1 in 5 or 0.20. Use the rules of probability to determine the likelihood that she will sell a policy to 3 of the 4 prospective clients.

Solution 2 – Using the Rules for a Binomial Probability Distribution

First note that the situation described meets the requirements of the binomial probability distribution. The conditions are:

1. There are a fixed number of trials—the representative visits four customers.
2. There are only two possible outcomes for each trial—she sells a policy or she does not sell a policy.
3. The probability of a success remains constant from trial to trial—for each appointment the probability of selling a policy (a success) is 0.20.
4. The trials are independent—if she sells a policy to the second appointment this does not alter the likelihood of selling to the third or the fourth appointment.

If *S* represents the outcome of a sale and *NS* the outcome of no sale, one possibility is that no sale is made on the first appointment but sales are made at the last 3.

(NS, S, S, S)

These events are independent; therefore the probability of their joint occurrence is the product of the individual probabilities. Therefore, the likelihood of no sale followed by three sales is (0.8) (0.2) (0.2) (0.2) = 0.0064. However, the requirements of the problem do not stipulate the location of NS. It could be the result of any one of the four appointments. The following summarizes the possible outcomes.

Location of *NS*	Order of Occurrence	Probability of Occurrence
1	*NS, S, S, S*	(0.8)(0.2)(0.2)(0.2) = 0.0064
2	*S, NS, S, S*	(0.2)(0.8)(0.2)(0.2) = 0.0064
3	*S, S, NS, S*	(0.2)(0.2)(0.8)(0.2) = 0.0064
4	*S, S, S, NS*	(0.2)(0.2)(0.2)(0.8) = 0.0064
Total		0.0256

The probability of exactly three sales in the four appointments is the sum of the 4 possibilities. Hence, the probability of selling insurance to 3 out of 4 appointments is 0.0256.

Example 3 – Using the Formula for a Binomial Probability Distribution

Now let's use formula [6-3] for the binomial distribution to compute the probability that the sales representative in Example 2 will sell a policy to exactly 3 out of the 4 prospective clients.

Solution 3 – Using the Formula for a Binomial Probability Distribution

To repeat, formula [6-3] for the binomial probability distribution is:

$$P(x) = {}_nC_x(\pi)^x(1-\pi)^{n-x} \qquad [6-3]$$

Where:

$_nC_x$ denotes a combination of n items selected x at a time
x is the number of successes, 3 in the example.
n is the number of trials, 4 in the example.
π is the probability of a success, 0.20.
$(1-\pi)$ is the probability of a failure, 0.80 found by (1 – 0.20).

The formula is applied to find the probability of selling an insurance policy to exactly 3 out of 4 potential customers.

$$P(x) = {}_nC_x(\pi)^x(1-\pi)^{n-x} = \frac{n!}{x!(n-x)!}(\pi)^x(1-\pi)^{n-x}$$

$$= \frac{4!}{3!(4-3)!}(0.20)^3(0.80)^{4-3} = 0.0256$$

Thus the probability is 0.0256 that the representative will be able to sell policies to exactly 3 out of the 4 clients visited. This is the same probability as computed earlier. Clearly, formula [6-3] leads more directly to a solution and better accommodates the situation where the number of trials is large.

Self-Review 6.2

Check your answers against those in the ANSWER section.

It is known that 60 percent of all registered voters in the 42nd Congressional District are Republicans. Three registered voters are selected at random from the district. Compute the probability that exactly 2 of the 3 selected are Republicans, using:

a. The rules of probability **b.** The binomial probability distribution formula.

Example 4 – Using Binomial Table to Determine Probability

In Examples 2 and 3 the probability of 3 sales resulting from 4 appointments was computed using both the rules of addition and multiplication and the binomial formula. A more convenient way of arriving at the probabilities for 0, 1, 2, 3, or 4 sales out of 4 appointments is to refer to a binomial table. We use the binomial table to determine the probabilities for all possible outcomes.

Solution 4 – Using Binomial Table to Determine Probability

Refer to Appendix A, the binomial table. Find the table where n, the number of trials, is 4. Next, find the row where $x = 0$, and move horizontally to the column headed $\pi = 0.20$. The probability of 0 sales is 0.410. The list for all possible outcome number of successes is shown at the right.

Binomial Probability Distribution	
$n = 4$	$\pi = 0.20$
Number of Successes (x)	Probability
0	0.410
1	0.410
2	0.154
3	0.026
4	0.002
	*1.002

*Slight discrepancy due to rounding

Example 5 – Determining the Mean of a Binomial Distribution

Use the information regarding the insurance representative, where $n = 4$ and $\pi = 0.20$, to compute the probability that the representative sells more than two policies. Also determine the mean and variance of the number of policyholders.

Solution 5 – Determining the Mean of a Binomial Distribution

The binomial table (Appendix A) can be used to determine the probability. First, note that the solution must include the probability that exactly 3 policies are sold and exactly 4 policies are sold, but not 2. From Appendix A, $P(3) = 0.026$ and $P(4) = 0.002$. The rule of addition is then used to combine these mutually exclusive events.

$$\begin{aligned} P(\text{more than 2}) &= P(3) + P(4) \\ &= 0.026 + 0.002 \\ &= 0.028 \end{aligned}$$

Thus the probability that a representative sells more than 2 policies is 0.028.

Suppose the question asked is: "What is the probability of selling three or more policies in four trials?" Since there are no outcomes between "greater than 2" and "less than 3", the answer is exactly the same (0.028).

To determine the mean of a binomial we use text formula [6-4].

$$m = n\pi = 4(0.20) = 0.80$$

Thus if the sales representative has several days with 4 appointments, typically he/she will sell 0.80 policies per day.

To determine the variance of a binomial we use text formula [6-5].

$$\sigma^2 = n\pi(1-\pi) = 4(0.20)(0.80) = 0.64$$

So the standard deviation is $\sqrt{0.64} = 0.80$.

Self-Review 6.3

Check your answers against those in the ANSWER section.

Labor negotiators estimate that 30 percent of all major contract negotiations result in a strike. During the next year, 12 major contracts must be negotiated. Determine the following probabilities using Appendix A:
a. no major strikes **b.** at least 5 **c.** between 2 and 4 (that is 2, 3, or 4).

Example 6 – Using the Poisson Probability Distribution

Alden & Associates write weekend trip insurance at a very nominal charge. Records show that the probability a motorist will have an accident during the weekend and will file a claim is quite small (0.0005). Suppose Alden wrote 400 policies for the forthcoming weekend. Compute the probability that exactly two claims will be filed. Depict this distribution in the form of a chart.

Solution 6 – Using the Poisson Probability Distribution

The Poisson distribution is appropriate for this problem because the probability of filing a claim is small (π = 0.0005), and the number of trials n is large (400).

The Poisson distribution is described by formula [6-6]:

$$P(x) = \frac{\mu^x e^{-\mu}}{x!} \qquad [6-6]$$

Where:
x is the number of successes (claims filed). In this example $x = 2$
μ is the expected or mean number of claims to be filed $\mu = n\pi =$ (400) (0.0005) = 0.2
e is a mathematical constant equal to 2.71828.

The probability that exactly two claims are filed is 0.0164, found by

$$P(2) = \frac{\mu^x e^{-\mu}}{x!} = \frac{(0.2)^2(2.718)^{-0.2}}{2!} = \frac{(0.04)(0.81874)}{(2)(1)} = 0.0164$$

Poisson Probability Distribution μ = 0.2	
Number of Claims	Probability
0	0.8187
1	0.1637
2	0.0164
3	0.0011
4	0.0001
	1.0000

This indicates that the probability is small (about 0.0164) that exactly 2 claims will be filed. The calculations to determine the probability 0.0164 were shown above. As noted previously, a convenient way to determine Poisson probabilities is to refer to Appendix C. To use this table, first find the column where $\mu = 0.20$, then go down that column to the row where $X = 2$ and read the value at the intersection. It is 0.0164.

The probabilities computed using formula [6-6] and those in Appendix C are the same. The complete Poisson distribution is shown at the right and a graph for the case where $\mu = 0.20$ is shown below. Note the shape of the graph. It is positively skewed, and as the number of claims increases, the probability of a claim decreases.

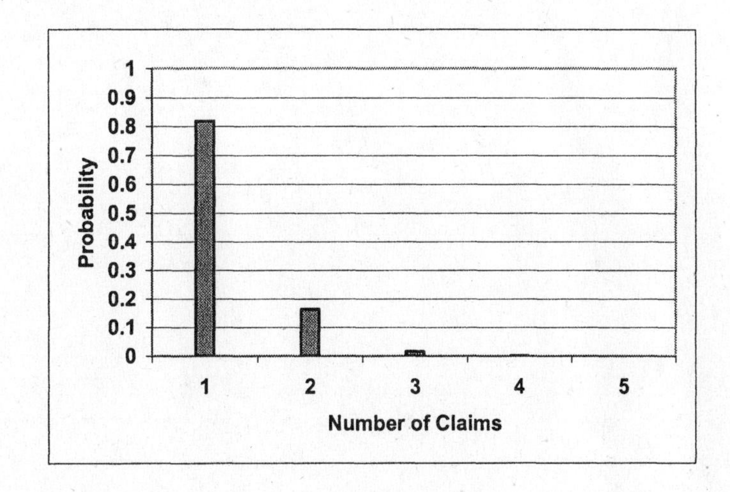

Self-Review 6.4

Check your answers against those in the ANSWER section.

The probability of a typographical error on any page is 0.002. If a textbook contains 1,000 pages, compute the probability there are:

a. No typos on a page

b. At least 2 typos on a page.

CHAPTER 6 ASSIGNMENT

DISCRETE PROBABILITY DISTRIBUTIONS

Name ______________________________ Section ________ Score ______

Part I Select the correct answer and write the appropriate letter in the space provided.

______ 1. A listing of all possible outcomes of an experiment and the corresponding probability is called:
- **a.** a random variable.
- **b.** a probability distribution.
- **c.** the normal rule.
- **d.** the complement rule.

______ 2. A probability distribution that can assume only certain values within a range is called
- **a.** a Poisson probability distribution.
- **b.** a continuous probability distribution.
- **c.** a random variable.
- **d.** a discrete probability distribution

______ 3. Which of the following is **not** a requirement of the binomial distribution?
- **a.** the trials must be independent
- **b.** the probability of a success changes from one trial to the next
- **c.** the sample size must be fixed
- **d.** only two outcomes are possible

______ 4. The mean of a discrete probability distribution is also called the
- **a.** variance.
- **b.** expected value.
- **c.** standard deviation.
- **d.** median.

______ 5. Which of the following statements is true about a Poisson probability distribution?
- **a.** The outcome of one trial affects the outcome of another trial.
- **b.** The sample size is small.
- **c.** Probability changes after each trial.
- **d.** The probability of a success is usually small.

______ 6. Which distribution would be most appropriate if one wanted to find the probability of selecting three Republicans from a sample of 15 politicians?
- **a.** binomial
- **b.** continuous
- **c.** normal
- **d.** Poisson

______ 7. A discrete distribution is usually the result of
- **a.** a measurement.
- **b.** a count.
- **c.** a small sample.
- **d.** a small probability.

______ 8. Which of the following is **not** a requirement for a discrete probability distribution?
- **a.** The trials are independent.
- **b.** The probability of each outcome is between 0 and 1.00.
- **c.** The outcomes are mutually exclusive.
- **d.** The sum of the probabilities is equal to 1.00.

______ 9. To construct a binomial probability distribution, we need to know
- **a.** the mean and standard deviation.
- **b.** only the mean.
- **c.** the size of the sample.
- **d.** the number of trials and probability of success.

Part II

Answer the questions by using the appropriate distribution. Show all your work. Write your answer in the answer box provided.

Number of times	Proportion
0	0.1
1	0.2
2	0.1
3	0.4
4	0.2

10. The number of connections on the Internet during any two-minute period is given by the distribution on the right.

a. Determine the mean number of times a connection is made during a two-minute period.

a.

b. Determine the standard deviation of the number of connections made during a two-minute period.

b.

11. According to a recent survey, 75% of all customers will return to the same grocery store. Suppose eight customers are selected at random, what is the probability that:

a. exactly five of the customers will return?

a.

b. all eight will return?

b.

c. at least seven will return?

c.

d. at least one will return?

d.

e. How many customers would be expected to return to the same store?

e.

12. Eighty percent of trees planted by a woodlands conservation group survive. What is the probability that:

a. 10 of the 12 trees just planted will survive?

a.

b. at least 10 of the trees just planted will survive?

b.

13. Customers use an automatic teller machine at an average rate of 15 per hour. What is the probability that exactly 12 will use the machine in the next hour?

14.

14. On the average two new checking accounts are opened per day at the Farmer's Bank. What is the likelihood that for a particular day:

a. no new accounts are opened?

a.

b. at least one new account is opened?

b.

15. A management team is comprised of six sales managers and four floor employees. A subcommittee of four is being formed to handle labor negotiations. What is the probability that two sales managers and two floor employees are selected?

16.

CHAPTER 7
CONTINUOUS PROBABILITY DISTRIBUTIONS

Chapter Goals

After completing this chapter, you will be able to:

1. Understand the difference between discrete and continuous distributions.
2. Compute the mean and standard deviation for a ***uniform distribution***.
3. Compute probabilities using the uniform distribution.
4. List the characteristics of a normal probability distribution.
5. Define and calculate ***z values***.
6. Determine the probability an observation is between two points on a normal distribution using the standard normal distribution.
7. Determine the probability that an observation is above (or below) a point on a normal distribution using the standard normal distribution.

Introduction

The previous chapter dealt with discrete probability distributions. Recall for a discrete distribution the outcome can assume only a specific set of values. For example, the number of correct responses to ten true-false questions can only be the numbers 0, 1, 2, ..., 10.

This chapter continues our study of probability distributions by examining the *continuous probability distribution*. Recall that a continuous probability distribution can assume an infinite number of values within a given range. As an example, the weights for a sample of small engine blocks are: 54.3, 52.7, 53.1 and 53.9 pounds.

We consider two families of continuous probability distributions, the ***uniform probability distribution*** and the ***normal probability distribution***. Both of these distributions describe the likelihood of a continuous random variable that has an infinite number of possible values within a specified range.

An example of a uniform probability distribution is the flight time between Detroit and Chicago. Suppose the time to fly from Detroit to Chicago is uniformly distributed within a range of 55 minutes to 75 minutes. We can determine the probability that we can fly from Detroit to Chicago in less than 60 minutes. Flight time is measured on a continuous scale.

The normal probability distribution is described by its mean and standard deviation. Suppose the life of an automobile battery follows the normal distribution with a mean of 36 months and a standard deviation of 3 months. We can determine the probability that a battery will last between 36 and forty months. Life of a battery is measured on a continuous scale.

The Family of Uniform Probability Distributions

The uniform probability distribution is the simplest distribution for a continuous random variable.

> ***Uniform probability distribution***: A continuous probability distribution with its values spread evenly over a range of values that are rectangular in shape and are defined by minimum and maximum values.

A uniform distribution is shown in Chart 7-1 from the textbook. The distribution's shape is rectangular and has a minimum value of "*a*" and a maximum value of "*b*". The height of the distribution is uniform for all values between "*a* " and "*b*". This implies that all the values in the range are equally likely.

Chart 7-1

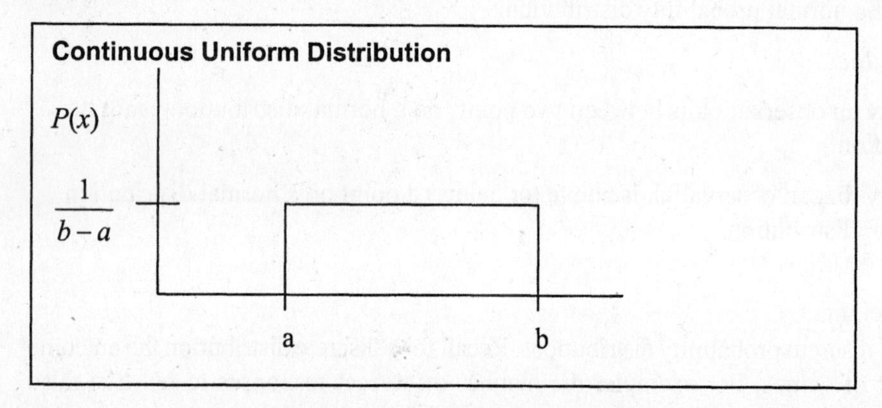

The mean of a uniform distribution is located in the middle of the interval between the minimum value of "*a* " and a maximum value of "*b*". It is calculated using formula [7–1]:

Mean of a Uniform Distribution	$\mu = \frac{a+b}{2}$	[7–1]

For example: Suppose that the time to fly from Detroit to Chicago is uniformly distributed within a range of 55 minutes minimum to 75 minutes maximum. The mean is found by using formula [7–1]:

$$\mu = \frac{a+b}{2} = \frac{55+75}{2} = \frac{130}{2} = 65$$

Thus the mean flight time is 65 minutes.

The standard deviation describes the dispersion of a distribution. In a uniform distribution, the standard deviation is also related to the interval between the minimum value of "*a* " and a maximum value of "*b*". It is calculated using formula [7–2]:

Standard Deviation of a Uniform Distribution	$\sigma = \sqrt{\frac{(b-a)^2}{12}}$	[7–2]

For the flight time from Detroit to Chicago example the standard deviation is calculated using formula [7–2]

$$\sigma=\sqrt{\frac{(b-a)^2}{12}}=\sqrt{\frac{(75-55)^2}{12}}=\sqrt{\frac{(20)^2}{12}}=\sqrt{\frac{400}{12}}=\sqrt{33.3333}=5.77=5.8$$

Thus the standard deviation for the flight is 5.8 minutes.

Another key element of the uniform distribution is the height: *P(x)*. The height is the same for all values of the random variable "*x* ".It is calculated using formula [7–3]:

Height of a Uniform Distribution	$P(x)=\frac{1}{(b-a)}$ if $a\le x\le b$ and 0 elsewhere	[7–3]

In Chapter 6, we discussed the fact that probability distributions are useful when making probability statements concerning the values of a random variable. Also for continuous random variables, areas within the distribution represent probabilities. Recall that: $\sum P(x)=1$ and $0\le P(x)\le 1$ for all values of x.

The relationship between area and probabilities is applied to the uniform distribution and its rectangular shape using the area of a rectangle formula. Recall that:

$$\text{Area of a rectangle} = \text{Height} \times \text{Base}$$

For a uniform distribution the height is $P(x)=\frac{1}{(b-a)}$ and the length is $(b-a)$. If we calculate the area of the rectangle we have:

$$\text{Area of a rectangle} = \text{Base} \times \text{Height}$$
$$=\frac{1}{(b-a)}\times(b-a)=1$$

Thus for any uniform distribution, the area under the curve is always 1.

For the flight time from Detroit to Chicago example the area is:

$$\text{Area of a rectangle} = \text{Height} \times \text{Base}$$
$$=\frac{1}{(b-a)}\times(b-a)=\frac{1}{(75-55)}\times(75-55)=\frac{20}{20}=1$$

The Family of Normal Probability Distribution

The Greek letter μ (lower case mu), represents the mean of a normal distribution and the Greek letter σ (lower case sigma) represents the standard deviation.

> ***Normal probability distribution***: A continuous probability distribution uniquely determined by μ and σ.

The major characteristics of the normal distribution are:

1. The normal distribution is "*bell-shaped*" and the mean, median, and mode are all equal and are located in the center of the distribution. Exactly one-half of the area under the normal curve is above the center and one-half of the area is below the center.

2. The distribution is *symmetrical* about the mean. A vertical line drawn at the mean divides the distribution into two equal halves and these halves possess exactly the same shape.

3. It is *asymptotic*. That is, the tails of the curve approach the *X*-axis but never actually touch it.

4. A normal distribution is completely described by its mean and standard deviation. This indicates that if the mean and standard deviation are known, a normal distribution can be constructed and its curve drawn.

5. There is a "family" of normal probability distributions. This means there is a different normal distribution for each combination of μ and σ.

These characteristics are summarized in the graph.

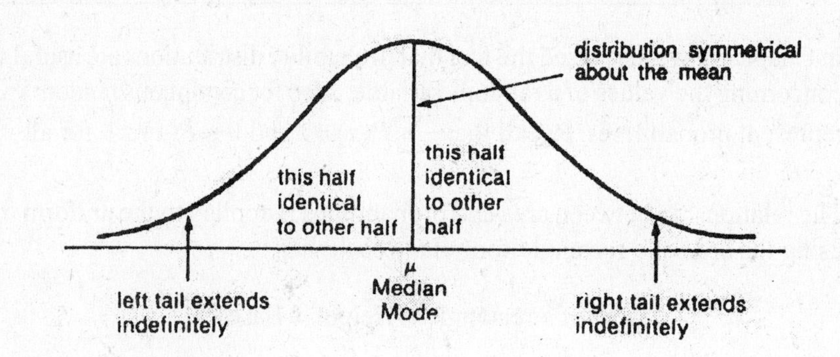

The Standard Normal Probability Distribution

As noted in the previous discussion, there are many normal probability distributions — a different one for each pair of values for a mean and a standard deviation. This principle makes the normal probability distribution applicable to a wide range of real-world situations. However, since there are an infinite number of probability distributions, it would be awkward to construct tables of probabilities for so many different normal distributions. An efficient method for overcoming this difficulty is to ***standardize*** each ***normal distribution***.

> ***Standard normal distribution***: A normal distribution with a mean of 0 and a standard deviation of 1.

An actual distribution is converted to a standard normal distribution using a ***z* value**.

> ***z value***: The signed distance between a selected value designated *X*, and the mean, μ, divided by the population standard deviation, σ.

The formula for a specific standardized *z* value is text formula [7–5]:

Standard Normal Value $$z=\frac{X-\mu}{\sigma} \quad [7-5]$$

Where:

X is the value of any particular observation or measurement.

μ is the mean of the distribution.

σ is the standard deviation of the distribution.

z is the standardized normal value, usually called the *z* value.

Applications of the Standard Normal Distribution

To obtain the probability of a value falling in the interval between the variable of interest (X) and the mean (μ), we first compute the distance between the value (X) and the mean (μ). Then we express that difference in units of the standard deviation by dividing ($X - \mu$) by the standard deviation. This process is called **standardizing.**

To illustrate the probability of a value being between a selected X value and the mean μ, suppose the mean useful life of a car battery is 36 months, with a standard deviation of 3 months. What is the probability that such a battery will last between 36 and 40 months?

The first step is to convert the 40 months to an equivalent standard normal value, using formula [7–5]. The computation is: $z = \dfrac{X - \mu}{\sigma} = \dfrac{40 - 36}{3} = \dfrac{4}{3} = 1.33$

Next refer to Appendix D, a table for the areas under the normal curve. A part of the table in Appendix D is shown at the right.

z	0.00	0.01	0.02	0.03	0.04	0.05
		•	•	•	•	
		•	•	•	•	
		•	•	•	•	
1.0						
1.1		0.3665	0.3686	0.3708	0.3729	
1.2		0.3869	0.3888	0.3907	0.3925	
1.3		0.4049	0.4066	0.4082	0.4099	
1.4		0.4207	0.4222	0.4236	0.4251	

To use the table, the z value of 1.33 is split into two parts, 1.3 and 0.03. To obtain the probability go down the left-hand column to 1.3, then move over to the column headed 0.03 and read the probability. It is 0.4082.

The probability that a battery will last between 36 and 40 months is 0.4082. Other probabilities may be calculated, such as more than 46 months, and less than 33 months. Further details are given in Examples 1 through 5.

Empirical Rule

Before examining various applications of the standard normal probability distribution, three areas under the normal curve will be considered which will be used in the following chapters. They were also called the Empirical Rule in Chapter 3.

1. About 68 percent of the area under the normal curve is within plus one and minus one standard deviation of the mean. This can be written as $\mu \pm 1\sigma$.
2. About 95 percent of the area under the normal curve is within plus and minus two standard deviations of the mean, written $\mu \pm 2\sigma$
3. Practically all of the area under the normal curve is within three standard deviations of the mean, written $\mu \pm 3\sigma$.

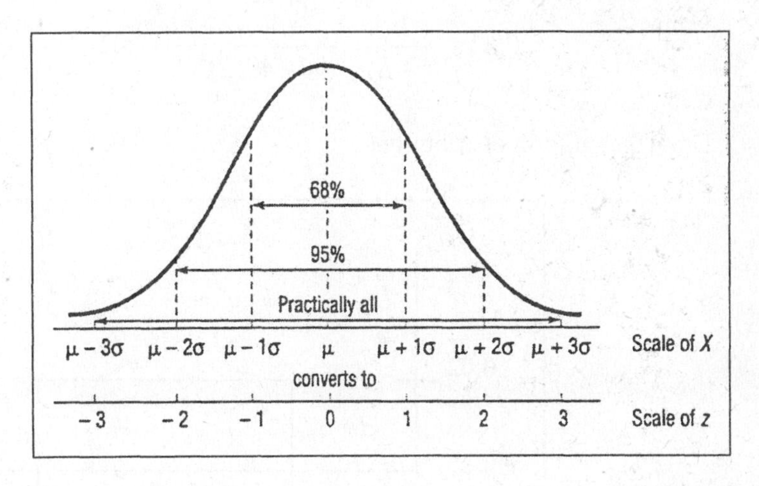

The estimates given above are the same as those shown on the diagram.

Glossary

Normal approximation to the binomial: A binomial probability can be estimated using the normal distribution.

Normal probability distribution: A continuous probability distribution uniquely determined by μ and σ.

Standard normal distribution: A normal distribution with a mean of 0 and a standard deviation of 1.

Uniform probability distribution: A continuous probability distribution with its values spread evenly over a range of values that are rectangular in shape and are defined by minimum and maximum values.

z value: The signed distance between a selected value designated X, and the mean, μ, divided by the population standard deviation, σ.

CHAPTER EXAMPLES

Example 1 –The Uniform Distribution

A group of statistics students collected data at the Cedar Point amusement park and reported that the "wait time" to ride the "Top Thrill Dragster" roller coaster is uniformly distributed within a range of 40 minutes to 90 minutes.

a. Determine the height and draw this uniform distribution.

b. Show that the total area under the curve is 1.00.

c. Determine the mean and standard deviation.

d. What is the probability a particular student will wait between 50 and 60 minutes?

e. What is the probability a particular student will wait less than 60 minutes?

Solution 1 – The Uniform Distribution

a. Determine the height using formula [7 – 3]

$$P(x) = \frac{1}{(b-a)} = \frac{1}{(90-40)} = \frac{1}{50} = 0.02$$

Draw the uniform distribution.

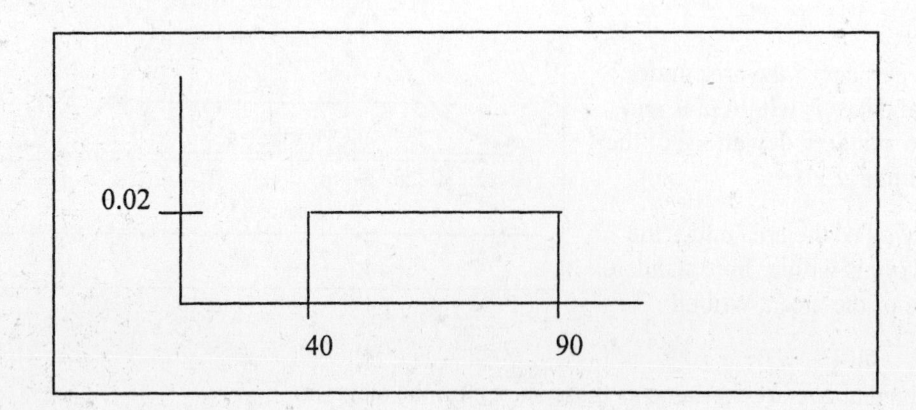

b. To show that the total area under the curve is 1.00, we use the area formula as follows:

$$\text{Area of a rectangle} = \text{Height} \times \text{Base}$$

$$= \frac{1}{(b-a)} \times (b-a) = \frac{1}{(90-40)} \times (90-40) = \frac{50}{50} = 1$$

c. To determine the mean use formula [7–1]. To determine the standard deviation use formula [7–2].

$$\mu = \frac{a+b}{2} = \frac{90+40}{2} = \frac{130}{2} = 65$$

$$\sigma = \sqrt{\frac{(b-a)^2}{12}} = \sqrt{\frac{(90-40)^2}{12}} = \sqrt{\frac{(50)^2}{12}} = \sqrt{208.333} = 14.43$$

d. The probability a particular student will wait between 50 and 60 minutes is found by finding the area of the rectangle with a height of 0.02 and a base of 10.

$$P(50 < \textit{wait time} < 60) = \textit{Height} \times \textit{Base} = \frac{1}{(90-40)} \times (60-50) = 0.02 \times 10 = 0.20$$

e. The probability a particular student will wait less than 60 minutes is found by finding the area of the rectangle with a height of 0.02 and a base of (60 – 40).

$$P(40 < \textit{wait time} < 60) = \textit{Height} \times \textit{Base} = \frac{1}{(90-40)} \times (60-40) = 0.02 \times 20 = 0.40$$

This probability is illustrated by the following graph.

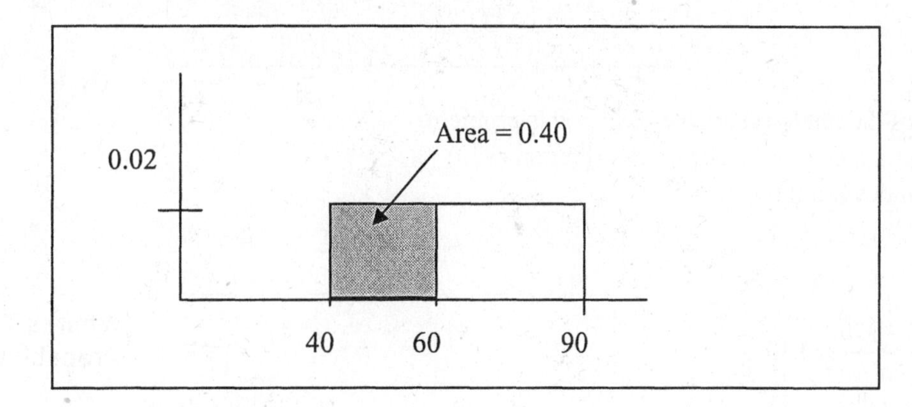

Self-Review 7.1

Check your answers against those in the ANSWER section.

The time that the customers at the "self serve" check out stations at the Mejers store spend checking out follows a uniform distribution between 0 and 3 minutes.

a. Determine the height and draw this uniform distribution.

b. How long does the typical customer wait to check out?

c. Determine the standard deviation of the wait time.

d. What is the probability a particular customer will wait less than one minute?

e. What is the probability a particular customer will wait between 1.5 and 2 minutes?

Example 2 – Solving for z and Areas Under the Curve

The mean amount of gasoline and services charged by Key Refining Company credit customers is $70 per month. The distribution of amounts spent is approximately normal with a standard deviation of $10. Compute the probability of selecting a credit card customer at random and finding the customer charged between $70 and $83 last month.

Solution 2 – Solving for z and Areas Under the Curve

The first step is to convert the area between $70 and $83 to a *z* value using formula [7-5].

$$z = \frac{X - \mu}{\sigma} \qquad [7-5]$$

Where:
X is any value of the random variable ($83 in this problem)
μ is the arithmetic mean of the normal distribution ($70)
σ is its standard deviation ($10).

Solving for z:

$$z = \frac{X - \mu}{\sigma} = \frac{\$83 - \$70}{10} = 1.30$$

This indicates that $83 is 1.30 standard deviations to the right of the mean of $70. Showing the problem graphically:

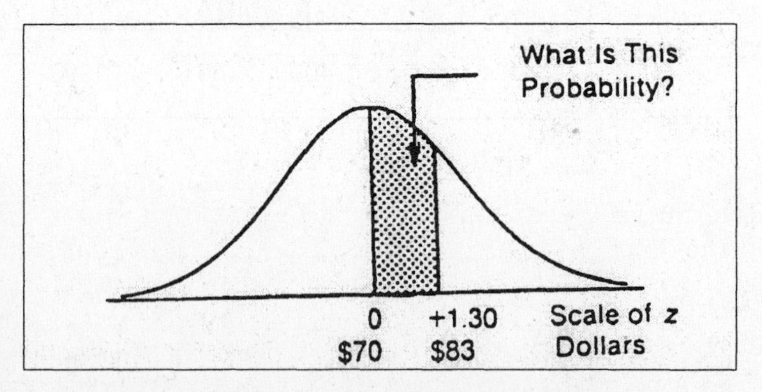

The probability of a *z* value from 0 to 1.30 is given in a table of areas of the normal curve, Appendix D. To obtain the probability, go down the left-hand column to 1.3, then move over to the column headed 0.00, and read the probability. It is 0.4032. To put it another way, 40.32 percent of the credit card customers charge between $70 and $83 per month.

Example 3 – Determining the Combined Probability and Areas Under the Curve

Again using the Key Refinery data from Example 2, compute the probability of customers charging between \$57 and \$83 per month.

Solution 3 – Determining the Combined Probability and Areas Under the Curve

As shown in the following graph, the probability of a customer charging between \$57 and \$70 per month must be combined with the probability of charging between \$70 and \$83 in a month to obtain the combined probability.

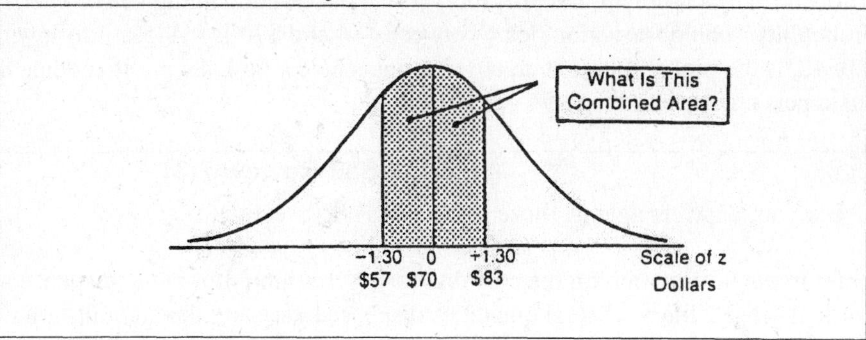

$$z = \frac{X - \mu}{\sigma} = \frac{\$57 - \$70}{\$10} = \frac{-13}{10} = -1.30$$

The probability of between \$70 and \$83 was computed in Example 2. Due to the symmetry of the normal distribution, the probability between 0 and 1.30 is the same as the probability between −1.30 and 0. It is 0.4032. The probability that customers will charge between \$57 and \$83 is 0.8064, found by adding 0.4032 and 0.4032.

Self-Review 7.2

Check your answers against those in the ANSWER section.

A cola-dispensing machine is set to dispense a mean of 2.02 liters into a bottle labeled 2 liters. Actual quantities dispensed vary and the amounts are normally distributed with a standard deviation of 0.015 liters.

a. What is the probability a bottle will contain between 2.02 and 2.04 liters?
b. What is the probability a bottle will contain between 2.00 and 2.03 liters?

Example 4 – Determining the *z* Probability

Using the Key Refining data from Example 2, what is the probability that a particular customer charges less than \$54?

Solution 4 – Determining the *z* Probability

The area to be determined is shown at the right.

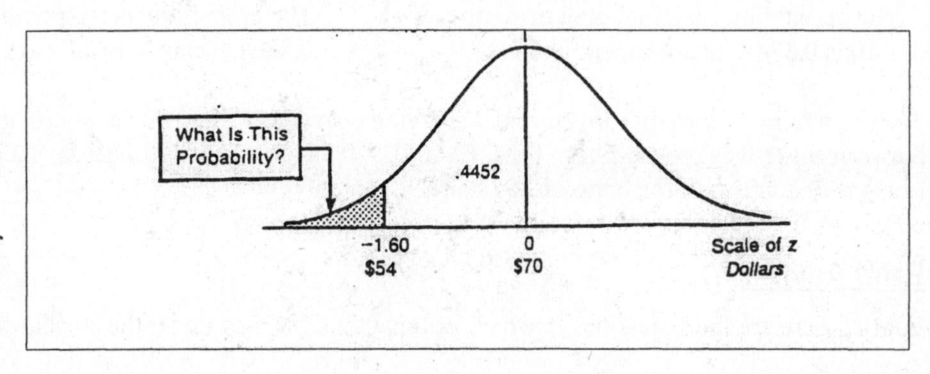

The *z* value for the area of the normal curve between \$70 and \$54 is −1.60, found by:

$$z = \frac{X - \mu}{\sigma} = \frac{\$54 - \$70}{\$10} = \frac{-\$16}{\$10} = -1.60$$

Referring to Appendix D, and a z of 1.60, the area of the normal curve between μ ($70) and X ($54) is 0.4452. Recall that for a symmetrical distribution half of the observations are above the mean, and half below it. In this problem, the probability of an observation being below $70 is, therefore, 0.5000. Since the probability of an observation being between $54 and $70 is 0.4452, it follows that 0.5000 – 0.4452 = 0.0548, is the probability that an observation is below $54. To put it another way, 5.48 percent of the customers charge less than $54 per month.

Self-Review 7.3

Check your answers against those in the ANSWER section.

Refer to the information on the cola-dispensing machine. It is set to dispense a mean of 2.02 liters into a bottle labeled 2 liters. Actual quantities dispensed vary and the amounts are normally distributed with a standard deviation of 0.015 liters. What is the probability a bottle will contain less than 2 liters?

Example 5 – Computing the Mean of a Probability Distribution

Again using the Key Refining data from Example 2, compute the probability of a customer charging between $82 and $92.

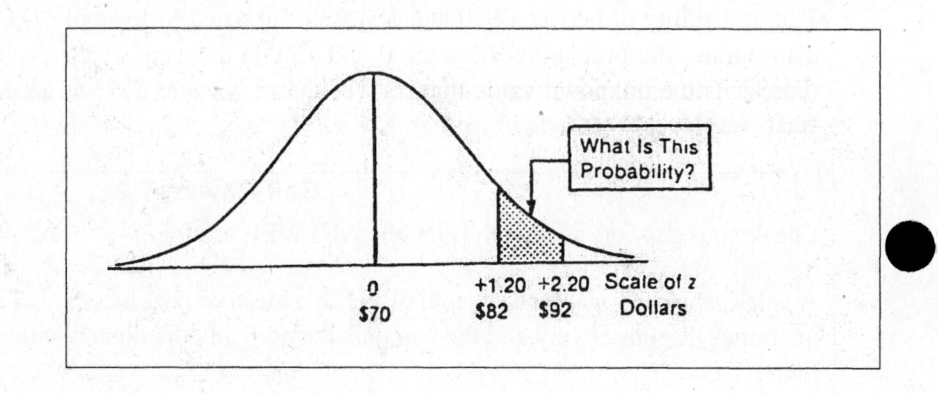

Solution 5 – Computing the Mean of a Probability Distribution

The area to be determined is depicted in the diagram.

The areas of the normal curve between $70 and $82, and $70 and $92 are determined using formula [7-5].

$$z = \frac{X - \mu}{\sigma} = \frac{\$82 - \$70}{\$10} = 1.20 \qquad z = \frac{X - \mu}{\sigma} = \frac{\$92 - \$70}{\$10} = 2.20$$

The probability corresponding to a z of 1.20 is 0.3849 (from Appendix D).

The probability corresponding to a z of 2.20 is 0.4861 (from Appendix D).

The probability of a credit card customer charging between $82 and $92 a month, therefore, is the difference between these two probabilities. Thus, (0.4861 – 0.3849) = 0.1012. That is, 10.12 percent of the charge account customers charge between $82 and $92 monthly.

Brief Review

In brief there are four situations in which you may find the area under the standard normal distribution.

1. To find the area between 0 and z or $(-z)$ you look up the value directly in the table. We did this in example 2.

2. To find the area between two points on different sides of the mean determine the two z values and add the corresponding areas. We did this in example 3.

3. To find the area beyond z or $(-z)$ locate the probability of z in the table and subtract that value from 0.500. We did this in example 4.

4. To find the area between two points on the same side of the mean, determine the two z values and subtract the smaller area from the larger area. We did this in example 5.

Example 6 – Computing Probability

Key Refining (Example 2) decided to send a special financing plan to charge account customers having the highest 10 percent of the money charges. What is the dividing point between the customers who receive the special plan and those who do not?

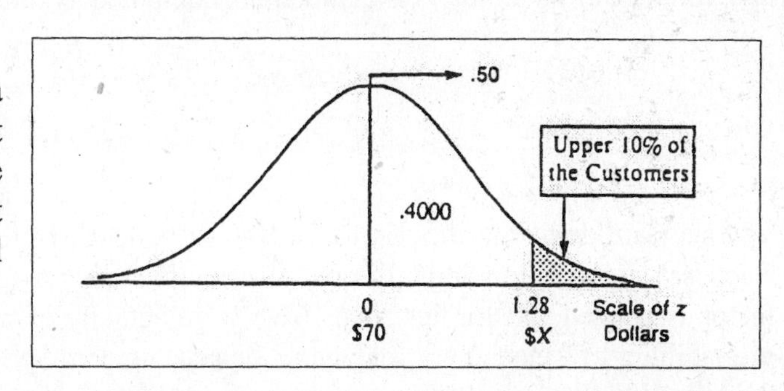

Solution 6 – Computing Probability

The shaded area in the diagram on the right represents the upper 10 percent who receive the special plan. X represents the unknown value that divides the customers into two groups—those who receive the special financing plan (the shaded area), and those who do not receive it. The area from the mean of $70 to this unknown X value is 0.4000, found by 0.5000 – 0.1000. From the table of areas of the normal curve (Appendix D), the closest z value corresponding to the area 0.4000 is 1.28. The area for $z = 1.28$ is 0.3997 and for $z = 1.29$ it is 0.4015. We select the z value as 1.28 because it is the closest to 0.4000. This indicates that the unknown X value is 1.28 standard deviations above the mean. Substituting 1.28 in the equation [7-5] for z:

$$z = \frac{X - \mu}{\sigma}$$

$$1.28 = \frac{X - \$70}{\$10}$$

$$X = \$82.80$$

Key Refining should send the special financing plan to those charge account customers having a monthly charge of $82.80 and above.

Self-Review 7.4

Check your answers against those in the ANSWER section.

Refer to the information on the cola-dispensing machine. It is set to dispense a mean of 2.02 liters into a bottle labeled 2 liters. Actual quantities dispensed vary and the amounts are normally distributed with a standard deviation of 0.015 liters. How much cola is dispensed in the largest 4% of the drinks?

Example 7 – Using Binomial Distribution

The Key Refining Company, referred to in the earlier problems, determined that 15 percent of its customers do not pay their bill by the due date. What is the probability that for a sample of 80 customers, less than 10 will not pay their bill by the due date?

Solution 7 – Using Binomial Distribution

The answer could be determined by using the binomial distribution where π, the probability of a success, is 0.15, and where n, the number of trials, is 80. However, most binomial tables do not go beyond an n of 25 and the calculations by hand would be very tedious.

As noted previously, the probability can be accurately estimated by using the normal approximation to the binomial. The approximations are quite good when both $n\pi$ and $n(1-\pi)$ are greater than 5.

In this case, $n\pi = (80)(0.15) = 12$, and $n(1-\pi) = (80)(1-0.15) = 68$. Both are greater than 5. Recall the mean and variance of a binomial distribution are computed as follows:

$$\mu = n\pi = (80)(0.15) = 12$$

$$\sigma^2 = n\pi(1-\pi) = (80)(0.15)(0.85) = 10.2$$

The standard deviation is 3.19, found by $\sqrt{10.2}$. The area less than 9.5 is shown on the following diagram. Because we are estimating a discrete distribution using a continuous distribution, the continuity correction factor is needed. In this instance if we were actually using the binomial distribution we would add the probabilities of 0 customers not paying, one customer not paying, and so on, up to nine customers not paying the bill. With the discrete distribution there would be no probability of 8.6 customers not paying their bill.

When we estimate binomial probabilities using the normal distribution, the area for nine corresponds to the area from 8.5 up to 9.5 and the area for 10 corresponds to 9.5 up to 10.5. In this case, we want all the area below (to the left of) 9.5. This area is depicted schematically as shown on the right:

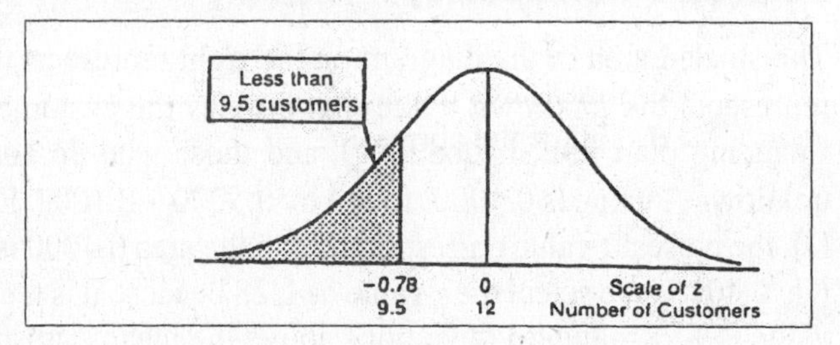

The z value associated with less than 9.5 customers is −0.78, found by

$$z = \frac{9.5 - 12.0}{3.19} = -0.78$$

The area to the left of −0.78 is 0.2177, found by (0.5000 − 0.2823). The probability that less than ten customers will not pay their bill is 0.2177.

Self-Review 7.5

Check your answers against those in the ANSWER section.

A new drug has been developed that is found to relieve nasal congestion in 90 percent of those with the condition. The new drug is administered to 300 patients with this condition. What is the probability that more than 265 patients will be relieved of the nasal congestion?

CHAPTER 7 ASSIGNMENT

CONTINUOUS PROBABILITY DISTRIBUTIONS

Name ______________________ Section __________ Score ______

Part I Select the correct answer and write the appropriate letter in the space provided.

______ 1. In a uniform distribution
- a. the mean and the median are always equal.
- b. the mean and the standard deviation are always equal.
- c. the mean is always larger than the median.
- d. the mean is always smaller than the median.

______ 2. The normal distribution is
- a. a bell-shaped distribution.
- b. a continuous distribution.
- c. symmetric.
- d. all of the above.

______ 3. The standard normal distribution
- a. is a special case of the normal distribution.
- b. has a mean equal to 0 and a standard deviation equal to 1.
- c. measures the distance from the mean in units of the standard deviation.
- d. all of the above.

______ 4. A normal distribution is completely described by
- a. its mean.
- b. its standard deviation.
- c. its mean and standard deviation.
- d. none of the above.

______ 5. Any normal distribution can be converted to a standard normal distribution by
- a. finding $\mu = n\pi$.
- b. determining that $n\pi$ is greater than 5.
- c. finding $z = \dfrac{X - \mu}{\sigma}$
- d. finding $\sigma = \sqrt{n\pi(1-\pi)}$

______ 6. A normal distribution
- a. has at least two peaks.
- b. is asymptotic.
- c. increases as X increases.
- d. is discrete.

______ 7. The normal distribution can be used to approximate the binomial when
- a. $n\pi$ is at least 25.
- b. both $n\pi$ and $n(1 - \pi)$ are greater than 5.
- c. $n\pi(1 - \pi p)$ is larger than 5.
- d. only when the z-score is above 5.

______ 8. What percent of the area under the normal curve is within $\mu \pm 1\sigma$?
- a. 68%.
- b. 34%.
- c. 95%.
- d. none of the above.

______ 9. A z value
- a. is the standard deviation for the standard normal distribution.
- b. is a measure of how many standard deviations the mean is from the median.
- c. is the difference of the mean and the probability of z.
- d. is a measure of how many standard deviations a particular score is from the mean.

Part II Answer the following questions. Be sure to show your work.

10. A statistics instructor collected data on the time it takes the students to complete a test. The test taking time is uniformly distributed within a range of 35 minutes to 55 minutes.

a. Determine the height and draw this uniform distribution.

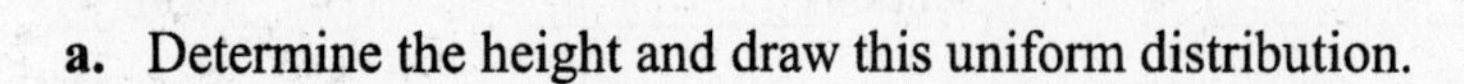

b. How long does the typical test taking time?

b.

c. Determine the standard deviation of the test taking time.

c.

d. What is the probability a particular student will take less than 45 minutes?

d.

e. What is the probability a particular student will take between 45 and 50 minutes?

e.

11. Fix-It Copiers advertises a mean time of 100 minutes for office calls with a standard deviation of 25 minutes. What percentage of calls are completed:

a. between 100 and 120 minutes?

a.

b. in less than 120 minutes?

b.

c. in less than 60 minutes?

c.

d. between 120 and 150 minutes?

d.

e. between 60 and 120 minutes?

e.

f. Twenty percent of their jobs take more than how much time?

f.

12. A certain printer cartridge has demonstrated a mean time usage of 9 hours and a standard deviation of 22 minutes. What time usage guarantee (in hours) should the manufacturer advertise in order to ensure that only 4% of the cartridges fail to meet the guaranteed time usage?

13.

13. A file cabinet manufacturer estimates that 5% of its file cabinets will have drawers that close improperly. Assume a production run of 120 cabinets is completed.

a. What is the mean and standard deviation of this distribution?

a.

b. What is the probability that 7 or more are defective?

b.

c. What is the probability that exactly 7 cabinets are defective?

c.

d. What is the probability that there are more than 8 defective file cabinets?

d.

14. A clothing store asserts that 60% of its customers pay by credit. On a particular day, 35 customers purchased items at the store.

a. What is the mean and standard deviation of the binomial distribution?

a.

b. What is the probability that half (18 or more) of the customers paid by credit?

b.

c. What is the probability that 30 or more paid by credit?

c.

d. What is the probability that less than 12 paid by credit?

d.

e. What is the probability that 12 to 30 paid by credit?

e.

15. It is very difficult for small businesses to be successful. The Small Business Administration estimates that 20 percent will dissolve or go bankrupt within two years. A sample of 50 new businesses is selected.

a. What is the mean and standard deviation of this distribution?

a.

b. What is the probability that more than 16 in the sample will go bankrupt?

b.

c. What is the probability that exactly 14 will go bankrupt?

c.

d. What is the probability that between 7 and 9 businesses will go bankrupt?

d.

e. What is the probability that between 7 and 15 businesses will go bankrupt?

e.

CHAPTER 8
SAMPLING METHODS AND THE CENTRAL LIMIT THEOREM

Chapter Goals

After completing this chapter, you will be able to:

1. Explain why a sample is often the only feasible way to learn something about a population.
2. Describe methods to select a sample.
3. Define and construct a sampling distribution of the sample mean.
4. Explain the ***central limit theorem***.
5. Use the ***central limit theorem*** to find probabilities of selecting possible sample means from a specified population.

Introduction

This chapter is the beginning of our study of sampling. Sampling is necessary because we want to make statements about a population but we do not want to (or cannot) examine all the items in that population. Recall from Chapter 1 that a **population** refers to the entire group of objects or persons of interest. The population of interest might be all the students enrolled in a business college or all the computer chips produced during the last hour. A **sample** is a portion, a part, or a subset of the population. Fifty college students out of 4,000 receiving payments might constitute the sample, or 20 computer chips might be sampled out of 1,500 produced last hour.

Reasons for Sampling

Why is it necessary to sample? Why can't we just inspect all the items? There are several reasons.

1. *The time to contact the whole population may be prohibitive.* To ask every eligible voter if they plan to vote for the current senator in the forthcoming election would take months. The electic would probably be over before the survey was completed.
2. *The cost of studying all the items in the population may be prohibitive.* Some television progr ratings are established by analyzing the viewing habits of about 1,200 viewers. The cost of studying all the homes having television would be exorbitant.
3. *The physical impossibility of checking all the items in the population.* The South Dakota Ga Commission, for example, cannot check all the deer, grouse, and other wild game because th always moving.
4. *The destructive nature of certain tests.* The manufacturer of fuses cannot test all of them because in the testing the fuse is destroyed and none would be available for sale.
5. *The sample results are adequate.* If the sample results of the viewing habits of 1,200 homes revealed that only 1.1 percent of the homes watched "60 Minutes," no doubt the program would be replaced by another show. Checking the viewing habits of all the homes regarding "60 Minutes" probably would not change the percent significantly.

Probability Sampling Methods

Four types of probability sampling are commonly used: ***simple random sampling***, ***systematic random sampling***, ***stratified random sampling***, and ***cluster sampling***.

Simple Random Sample

The most widely used type of sampling is a simple random sample.

> ***Simple random sample***: A sample selected so that each item or person in the population has the same chance of being included.

Several ways of selecting a simple random sample are:

1. The name or identifying number of each item in the population is recorded on a slip of paper and placed in a box. The slips of paper are shuffled and the required sample size is chosen from the box.
2. Each item is numbered and a *table of random numbers*, such as the one in Appendix E, is used to select the members of the sample.
3. There are many software programs, such as MINITAB and Excel, which have routines that will randomly select a given number of items from the population.

Systematic Random Sample

Another type of sampling is ***a systematic random sample***.

> ***Systematic random sample***: A random starting point is selected and then every k^{th} member of the population is selected.

In a systematic random sample the items or individuals of the population are arranged in some way — alphabetically, in a file drawer by date received, or by some other method. A random starting point is selected, and then every k^{th} member of the population is selected for the sample. In a systematic random sample, you might take all the items in the population and number them 1, 2, 3,.... Next, a random starting point is selected, let's say 39. Every k^{th} item thereafter, such as every 100th, is selected for the sample. This means that 39, 139, 239, 339, and so on would be a part of the sample.

Stratified Random Sample

Another type of probability sample is referred to as ***stratified random sampling***.

> ***Stratified random sample***: A population is divided into subgroups, called strata, and a sample is randomly selected from each stratum.

For example, if our study involved Army personnel, we might decide to stratify the population (all Army personnel) into generals, other officers, and enlisted personnel. The number selected from each of the three strata could be proportional to the total number in the population for the corresponding strata. Each member of the population can belong to only one of the strata. That is, a military person cannot be a general and a private at the same time.

Cluster Sampling

Another common type of sampling is ***cluster sampling***.

> ***Cluster sampling***: A population is divided into clusters using naturally occurring geographic or other boundaries. Clusters are then randomly selected and a sample is collected by randomly selecting from each cluster.

Cluster sampling is often used to reduce the cost of sampling when the population is scattered over a large geographic area. Suppose the objective is to study household waste collection in a large city.

Step 1: Divide the city into smaller units (perhaps precincts).

Step 2: The precincts are numbered and several selected randomly.

Step 3: Households within each of these precincts are randomly selected and interviewed.

Sampling "Error"

It is not logical to expect that the results obtained from a sample will coincide exactly with those from a population. For example, it is unlikely that the mean welfare payment for a sample of 50 recipients is exactly the same as the mean for all 4,000 welfare recipients. We expect a difference between a sample statistic and its corresponding population parameter. The difference is called ***sampling error***.

> ***Sampling error***: The difference between a sample statistic and its corresponding population parameter.

Because these errors happen by chance, they are referred to as chance variations.

Sampling Distribution of the Sample Mean

Suppose all possible samples of size n are selected from a specified population, and the mean of each of these samples is computed. The distribution of these sample means is called the ***sampling distribution of the sample mean.***

> ***Sampling distribution of the sample mean:*** A probability distribution of all possible sample means of a given sample size.

The sampling distribution of the mean is a probability distribution and has the following major characteristics:

1. The mean of all the sample means will be exactly equal to the population mean.
2. If the population from which the samples are drawn is normal, the distribution of sample means is also normally distributed.
3. If the population from which the samples are drawn is not normal, the sampling distribution is approximately normal, provided the samples are "sufficiently" large (usually accepted to include at least 30 observations).

The Central Limit Theorem

The ***central limit theorem*** states that, for large random samples, the shape of the sampling distribution of the sample means is close to a normal probability distribution. The approximation is more accurate for large samples than for small samples. We can make logical and reasonable statements about the distribution of the sample means with little or no information about the shape of the original distribution from which we took the sample.

This phenomenon is called the ***central limit theorem***.

> ***Central limit theorem*:** If all samples of a specified size are selected from any population, the sampling distribution of the sample mean is approximately a normal distribution. This approximation improves with larger samples.

Standard Error of the Mean

The Central Limit Theorem does not address the dispersion of the sampling distribution of sample means nor does it address the comparison of the sampling distribution of sample means to the mean of the population. It can be shown that the mean of the sampling distribution is the population mean, and if the standard deviation in the population is σ, the standard deviation of the means is $\frac{\sigma}{\sqrt{n}}$, where n is the number of observations in each sample. We refer to $\frac{\sigma}{\sqrt{n}}$ as the ***standard error of the mean.*** It is actually the standard deviation of the sampling distribution of the sample mean:

> ***Standard Error of the Mean*:** The standard deviation of the sampling distribution of the sample mean.

The standard error is a measure of the variability of the sampling distribution of the means. It is computed using text formula [8-1]

Standard Error of the Mean $$\sigma_{\bar{X}} = \frac{\sigma}{\sqrt{n}} \qquad [8-1]$$

Where:

$\sigma_{\bar{X}}$ is the standard error of the mean

σ is the population standard deviation

n is the sample size

In most situations we do not know the population standard deviation so we replace it with the sample standard deviation. We replace σ with s. Thus we have the following formula:

$$s_{\bar{X}} = \frac{s}{\sqrt{n}}$$

The size of the standard error is affected by the standard deviation. As the standard deviation increases so does the standard error. The standard error is also affected by the sample size. As the sample size increases the standard error decreases, which indicates that there is less variability in the distribution of the sample means. Obviously we conclude that as we increase the sample size the standard error decreases.

It is important to note the following:

1. The mean of the distribution of sample means will be *exactly* equal to the population mean if we are able to select all possible samples of a particular size from a given population. That is $\mu = \mu_{\overline{X}}$.
   Even if we do not select all samples, we can expect the mean of the distribution of the sample mean to be close to the population mean.
2. There will be less dispersion in the sampling distribution of the distribution of sample mean than in the population. If the standard deviation of the population is σ, the standard deviation of the distribution of sample means is $\frac{\sigma}{\sqrt{n}}$. Note that when we increase the size of the sample the standard error of the mean decreases.

Using the Sampling Distribution of the Sample Mean

The majority of statistical business decisions are made on the basis of sampling. Generally we have a population and wish to know something about that population, such as the mean. We take a sample from that population and wish to conclude whether the sampling error, that is the difference between the population parameter and the sample statistic, is due to chance.

We can compute the probability that a sample mean will fall within a certain range. The sampling distribution of the sample mean will follow the normal probability distribution under two conditions:

1. When the samples are taken from populations known to follow the normal distribution. In this case the size of the sample is not a factor.
2. When the shape of the population distribution is not known or the shape is known to be nonnormal, but the sample contains at least 30 observations.

Recall that we used the *z*-value found with formula [7-5] to convert any normal distribution to the standard normal distribution. We can use the standard normal table to find the probability of selecting a value of an observation that falls within a specified range. The formula is:

Standard Normal Value	$z = \frac{X - \mu}{\sigma}$	[7-5]

In this formula X is the value of the random variable, μ is the population mean, and σ is the population standard deviation.

Since most business decisions are based on a sample, we are interested in the distribution of the sample mean $\overline{X}$ not the value of X, the value of one observation. Formula [7-5] is altered to reflect this need. We change X to $\overline{X}$. Then we change the population standard deviation to the standard error of the mean: $\frac{\sigma}{\sqrt{n}}$.

Thus we have formula [8-2] that is used to find the *z* value for a normal population with a known population mean and standard deviation:

Finding the z Value of $\overline{X}$ When the Population Standard Deviation is Known

$$z = \frac{\overline{X} - \mu}{\sigma / \sqrt{n}} \quad [8-2]$$

If we do not know the value of the population standard deviation σ and the sample size is at least 30, we estimate the population standard deviation with the sample standard deviation s. Thus we use s to replace σ, the new formula is formula [8-3]:

Finding the z Value of $\overline{X}$ When the Population Standard Deviation is Unknown

$$z = \frac{\overline{X} - \mu}{s / \sqrt{n}} \quad [8-3]$$

Glossary

Central limit theorem: If all samples of a specified size are selected from any population, the sampling distribution of the sample means is approximately a normal distribution. This approximation improves with larger samples.

Cluster sampling: A population is divided into clusters using naturally occurring geographic or other boundaries. Clusters are then randomly selected and a sample is collected by randomly selecting from each cluster.

Sampling distribution of the sample mean: A probability distribution of all possible sample means of a given sample size.

Sampling error: The difference between a sample statistic and its corresponding population parameter.

Simple random sample: A sample selected so that each item or person in the population has the same chance of being included.

Standard Error of the Sample Mean: The standard deviation of the sampling distribution of the sample means.

Stratified random sample: A population is divided into groups, called strata, and a sample is randomly selected from each stratum.

Systematic random sample: A random starting point is selected and then every k^{th} member of the population is selected.

Chapter Examples

Example 1 – Selecting a Random Sample

Listed below are the advertisers in the "Coupon Section" of the Fort Walton Beach/Destin Florida phone directory. Also noted is whether the *type* of advertiser is a service advertiser (**S**), personal care advertiser (**P**), auto related advertiser (**A**), or recreation oriented advertiser (**R**). Some of the advertisers are to be randomly selected and asked various questions regarding the coupons.

00	A to Z Lock & Safe	S	10	Meineke Discount Mufflers	A
01	Action on the Blackwater River	R	11	Merry Maids	S
02	Atlas Exterminating	S	12	Midas Auto Service Experts	A
03	Blackwater Canoe Rental	R	13	Payless Mechanical, Inc.	S
04	Cain's AC & Refrigeration	S	14	Private Mini Storage	S
05	Clean and Fresh Carpet	S	15	Profast Auto Service Centers	A
06	Dale's Carpet Cleaning	S	16	Shalimar Point Golf Club	R
07	Executive Car Wash	A	17	Studio 21 Hair Designs	P
08	Florida Steam Carpet	S	18	Trees Unlimited	S
09	Fred Astaire Dance Studio	R	19	Tropical Images Salon	P

a. Randomly select a sample of five advertisers.
b. Randomly select a second sample of five advertisers.
c. Randomly select a sample that consists of every fourth advertiser.
d. Select a sample of four advertisers so that one of each type of advertiser is included.

Solution 1 – Selecting a Random Sample

a. Use Appendix E – Table of Random Numbers – and pick a starting point. The current time is 11:50 so we will start with the eleventh row down and the fifth column in from the left. The number is 07960.

- If we start with the first two digits, then the first two-digit number equal to or less than 19 is **07.**
- Working down the column the next two-digit number equal to or less than 19 is **05.**
- We skip 38, 88, 95, and pick **14.**
- We continue down the column and skip 76, 92, 46, and pick **03.**
- We skip 21, 89 and pick **06.**
- We have chosen advertisers number **07, 05, 14, 03,** and **06.** The random sample of advertisers is shown at the right.

03	Blackwater Canoe Rental
05	Clean and Fresh Carpet
06	Dale's Carpet Cleaning
07	Executive Car Wash
14	Private Mini Storage

b. Use Appendix E – Table of Random Numbers – and pick a starting point. The current time is again 11:50 so we will start with the eleventh row down and the fifth column in from the left. The number is 07960. This time we will use the last two digits in the column.

- If we start with the last two digits then we need to skip 60, 43, 33, and the first two-digit number equal to or less than 19 is **08.**

- Working down the column we skip 85, 36, 39, 34, 91, 85, 64, 45, 79, 22, 85, 64, and pick **14**.

- We continue down the column we skip 80, 99, 94, 87 and the next two digit number equal to or less than 19 is **12**.

- We skip 33, 61, 83, 22, and pick **06**.

- We continue down the column and skip 29, 32, 75, 86, 72, 46, 70, and pick **17**.

- We have chosen advertisers number **08, 14, 12, 06,** and **17**. The random sample of advertisers is shown at the right.

06	Dale's Carpet Cleaning
08	Florida Steam Carpet
12	Midas Auto Service Experts
14	Private Mini Storage
17	Studio 21 Hair Designs

It is interesting to note that advertiser 06 and 14 were picked for both samples.

c. In order to pick every fourth advertiser we need an initial starting point. We need to determine if we will start with advertiser 00, 01, 02 or 03. We use the random number table and simply close our eyes and place our finger on the table. We picked the array that is in the sixteenth (16) row and the third (3) column headed by the numbers 37722.

- We move down the first column until we come to a single digit number less than or equal to 03. Thus we skip 37, 43, and 92 and pick **01**.

- We will pick our sample starting with **01** and pick every fourth advertiser, which would be **05, 09, 13,** and **17**. The random sample of advertisers is shown at the right.

01	Action on the Blackwater River
05	Clean and Fresh Carpet
09	Fred Astaire Dance Studio
13	Payless Mechanical, Inc.
17	Studio 21 Hair Designs

d. The process is similar to what we did in parts **a** and **b,** with the exception that we have to pay attention to the type of service performed by each advertiser. Use Appendix E – Table of Random Numbers – and pick a starting point. The current time is 17:10 so we will start with the seventeenth row down and the tenth column in from the left. The number is 72949.

- This time we will use the second and third digits in the column. If we start with the second and third digits then we need to skip 29, 41, 56, 92, 78, and the first two-digit number less than or equal to 19 is **12**. We note that this is an auto care advertiser **(A)**.

- Working down the column, skip 69, 25, 48, and pick **17**. This is a personal care advertiser **(P)**.

- We continue down the column and skip 62. The next two-digit number equal to or less than 19 is **07**. We note that this is an auto service advertiser (A) and skip it since we already have such an advertiser.

- We skip 30, 61, 27, and pick **18**. We note that this is a service firm advertiser **(S)**.

- We skip 40, 59, 81, 25, and pick **15**. We skip this one since it is also auto related **(A)**.

- The next number is **12** which is also auto related **(A)**.

- We skip 59, 95, 76, 38, 22, 88, 20, 44, 25, and pick **16**. We note that this is a recreation advertiser(**R**). Thus we have selected an advertiser from each of the types. The random sample of advertisers is:

12	Midas Auto Service Experts (A)
17	Studio 21 Hair Designs (P)
18	Trees Unlimited (S)
16	Shalimar Point Golf Club (R)

Self-Review 8.1

Check your answers against those in the ANSWER section.

Refer to Example 1 and the advertisers in the "Coupon Section" of the Fort Walton Beach/Destin Florida phone directory. Use Appendix E – Table of Random Numbers.

a. Pick a sample of five advertisers. Start with row six and column six. The number is 84822. Use the first two numbers starting with 84.

b. Randomly select a second sample of five advertisers starting at the bottom right corner of the table. The number is 70603. Use the far right two digits starting with 03.

c. Randomly select a sample that consists of every fourth advertiser starting with 03.

d. Select a sample of four advertisers so that one of each type of advertiser is included. Start at the top of column two. The number is 08182. Use the left two digits starting with 08.

Example 2 – Developing a Probability Distribution

Suppose that a population consists of the six families living in Brentwood Circle. You are studying the number of children in the six families. The population information is shown at the right.

Family	Number of Children
Clark	1
Walston	2
Dodd	3
Marshall	5
Saner	3
White	4

List the possible samples of size 2 that could be selected from this population and compute the mean of each sample. Organize these sample means into a probability distribution.

Solution 2 – Developing a Probability Distribution

There are 15 different samples. The formula for the number of combinations is used to determine the total number of samples. Combination formula [5–9] is used.

$${}_nC_r = \frac{n!}{r!(n-r)!}$$

There are six members of the population and the sample size is two.

$${}_6C_2 = \frac{6!}{2!4!} = 15$$

Thus there are 15 different samples, as shown.

Sample Number	Families in the Sample	Total Number of Children in Sample	Mean Number of Children Per Family in Sample	
1	Clark, Walston	3	1.5	←3/2
2	Clark, Dodd	4	2.0	←4/2
3	Clark, Marshall	6	3.0	←6/2
4	Clark, Saner	4	2.0	←4/2
5	Clark, White	5	2.5	←5/2
6	Walston, Dodd	5	2.5	←5/2
7	Walston, Marshall	7	3.5	←7/2
8	Walston, Saner	5	2.5	←5/2
9	Walston, White	6	3.0	←6/2
10	Dodd, Marshall	8	4.0	←8/2
11	Dodd, Saner	6	3.0	←6/2
12	Dodd, White	7	3.5	←7/2
13	Marshall, Saner	8	4.0	←8/2
14	Marshall, White	9	4.5	←9/2
15	Saner, White	7	3.5	←7/2
		Total	45.0	

This information is organized into the following probability distribution called the sampling distribution of the means.

Mean Number of Children	Frequency	Probability	
1.5	1	0.067	← 1/15
2.0	2	0.133	← 2/15
2.5	3	0.200	← 3/15
3.0	3	0.200	← 3/15
3.5	3	0.200	← 3/15
4.0	2	0.133	← 2/15
4.5	1	0.067	← 1/15
	15	1.000	

Example 3 – Comparing the Means

Using the Brentwood Circle data in Example 2, compare the mean of the sampling distribution with the mean of the population. Compare the spread of the sample mean with that of the population.

Solution 3 – Comparing the Means

The mean of the sampling distribution and the mean of the population are the same. The population mean, written μ, is found by $\mu = [(1+2+3+5+3+4) \div 6] = 3.0$. The mean of the sampling distribution (written $\mu_{\bar{x}}$) because it is the mean of a group of sample means) is also 3.0, found by 45.0 ÷ 15.

$\bar{X}$ Sample Means	f Frequency	$f\bar{X}$
1.5	1	1.5
2.0	2	4.0
2.5	3	7.5
3.0	3	9.0
3.5	3	10.5
4.0	2	8.0
4.5	1	4.5
	15	45.0

The calculations for the sample means are:

$$\mu_{\bar{x}} = \frac{\Sigma f\bar{X}}{\Sigma f} = \frac{45.0}{15} = 3.0$$

The population mean μ is exactly equal to the mean of the sampling distribution $\mu_{\bar{x}}$ (3.0). This is always true.

Note in the graphs at the right that there is less spread in the sampling distribution of the means (bottom chart) than in the population distribution (top chart). The sample means range from 1.5 to 4.5, whereas the population values ranged from 1 to 5.

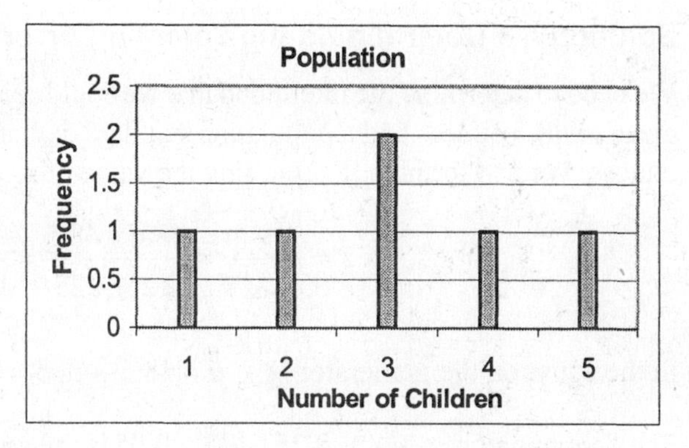

Also note that the shape of the population is different than that of the sampling distribution. This phenomenon is described by the central limit theorem. Recall the central limit theorem states that regardless of the shape of the population the sampling distribution will tend toward normal as n increases.

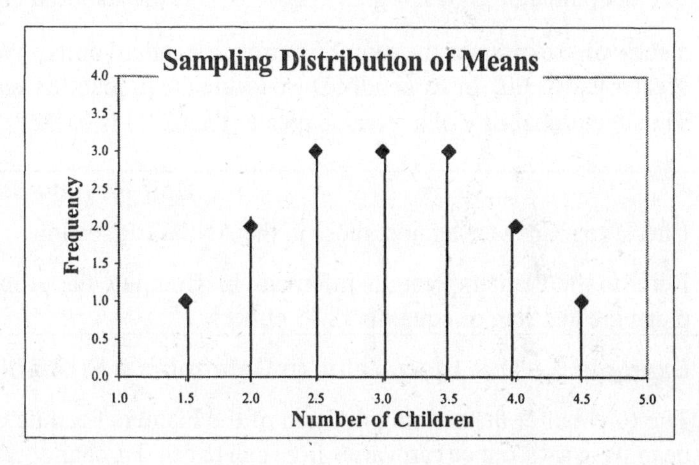

Self-Review 8.2

Check your answers against those in the ANSWER section.

The real estate company of Kuhlman and Associates has five sales people. Listed below is the number of homes sold last month by each of the five associates. Bruce Kuhlman, the owner, wants to estimate the population mean number of homes sold based on samples of three.

	Associate	# Sold
A.	Sue Klaus	6
B.	John Bardo	2
C.	Jean Cannon	5
D.	A.J. Kemper	9
E.	Carol Ford	3

a. If samples of size 3 are selected, how many different samples are possible?

b. List the various samples and compute the mean of each.

c. Develop a sampling distribution of the means.

d. Draw graphs to compare the variability of the sampling distribution of the mean with that of the population.

Example 4 – Determining the Sampling Error

The foreman of the canning division of the Planters Peanuts Company observed that the amount of cashews in a 48-ounce can varies from can to can. Records indicate that the process follows the normal probability distribution with a mean of 48.5 ounces and a standard deviation of 0.25 ounces. The foreman randomly selects 25 cans from the canning line and determines that the mean amount of cashews per can is 48.6 ounces. Compute the probability that the sample of 25 cans would have a mean greater than or equal to 48.6 ounces.

Solution 4 – Determining the Sampling Error

We need to determine the likelihood that we could select a sample of 25 cans from a normal population with a mean of 48.5 ounces and a population standard deviation of 0.25 ounces and find the sample mean to be 48.6 ounces. We use formula [8-2] to find the value of z.

$$z = \frac{\overline{X} - \mu}{\sigma / \sqrt{n}} = \frac{48.6 - 48.5}{0.25 / \sqrt{25}} = \frac{0.1}{0.05} = 2$$

In the equation the numerator $\overline{X} - \mu = 48.6 - 48.5 = 0.1$ is the sampling error.

The denominator $\sigma / \sqrt{n} = 0.25 / \sqrt{25} = 0.05$ is the standard error of the distribution of sample means.

The z-value expresses the sampling error in standard units. We need to compute the probability of a z-value greater than 2.00. In Appendix D we locate the probability corresponding to a z-value of 2.00. It is 0.4772. Thus the probability of a z-value greater than 2.00 is 0.0228, found by (0.5000 –0.4772) = 0.0228.

Self-Review 8.3

Check your answers against those in the ANSWER section.

Refer to the Planters Peanuts information. Compute the probability that a sample of 16 cans would have a mean greater than or equal to 48.45 ounces.

Example 5 – Use the z Value to Determine a Probability

The foreman of the canning division of the Planters Peanuts Company observed that the amount of redskin peanuts in a 48-ounce can varies from can to can. He cannot sample every can, but he knows that if he takes a sample of at least 30 cans the sampling distribution follows the normal distribution. The foreman randomly selects 100 cans from the canning line and determines that the mean amount of redskin peanuts per can is 48 ounces with a standard deviation of 0.45 ounce. What is the probability of finding a sample with a mean of 47.92 ounces or less from the population?

Solution 5 – Use the z Value to Determine a Probability

We assume that since the sample is sufficiently large that the sampling distribution of the sample mean follows the normal distribution. We use formula [8-3] to find the value of z.

$$z = \frac{\overline{X} - \mu}{s / \sqrt{n}} = \frac{47.92 - 48}{0.45 / \sqrt{100}} = \frac{-0.08}{0.045} = -1.77777 = -1.78$$

Referring to Appendix D, the z-value for – 1.78 is 0.4625. The likelihood of finding a z-value less than –1.78 is found by (0.5000 – 0.4625) = 0.0375. There is about a four percent chance that we could select a sample of 100 cans of redskin peanuts and find the mean of the sample is 47.92 ounces or less, when the population mean is 48 ounces.

Self-Review 8.4

Check your answers against those in the ANSWER section.

Refer to the Planters Peanuts information for redskin peanuts. A sample of 64 cans has a mean of 47.88 ounces and a standard deviation of 0.48 ounces. What is the probability of finding a sample with a mean of 47.88 ounces or less from the population?

CHAPTER 8 ASSIGNMENT

SAMPLING METHODS AND THE CENTRAL LIMIT THEOREM

Name ______________________________________ Section __________ Score ________

Part I Select the correct answer and write the appropriate letter in the space provided.

______ **1.** The *population proportion* is an example of a
a. sample statistic. **b.** normal population.
c. sample mean. **d.** population parameter.

______ **2.** In a *probability sample* each item in the population has
a. a chance of being selected.
b. the same chance of being selected.
c. a 50 percent chance of being selected.
d. no chance of being selected

______ **3.** In a *simple random sample* each item in the population has
a. a chance of being selected.
b. the same chance of being selected.
c. a 50 percent chance of being selected.
d. no chance of being selected..

______ **4.** The *sampling error* is
a. the difference between a sample statistic and a population parameter.
b. always positive.
c. the difference between the z value and the mean.
d. equal to the population value.

______ **5.** The *sample mean* is an example of a
a. sample statistic.
b. normal population.
c. weighted mean.
d. population parameter.

______ **6.** Suppose we have a negatively skewed population. According to the central limit theorem, the distribution of a sample mean of a particular size will
a. also be negatively skewed.
b. form a binomial distribution.
c. approach a normal distribution.
d. become positively skewed.

______ **7.** The population is the five employees in a physician's office. The number of possible samples of 2 that could be selected from this population is
a. 5 **b.** 10
c. 15 **d.** 60

_______8. The *sampling distribution of the sample mean* is the probability distribution of all
a. the sample statistics and their probability of occurrence..
b. the normal population parameters and their probability of occurrence..
c. the possible sample means and their probability of occurrence.
d. sample means.

_______9. In *cluster sampling*
a. each item in the population has a chance of being selected more than once.
b. the population is divided into primary units, and then samples are drawn from these units.
c. each item in a primary unit has a 50 percent chance of being selected.
d. every k^{th} item has a chance of being selected.

_______10. In a *systematic sample* a random starting point is chosen, and
a. each item in the population has a chance of being selected more than once.
b. the population is divided into primary units, and then samples are drawn from these units.
c. each item in a population has a 50 percent chance of being selected.
d. every k^{th} item thereafter is selected for the sample.

Part II Answer the following questions. Be sure to show essential work.

11. Listed below are the rental agencies in the *Automobile Rental* section in the phone directory. Also noted is whether the rental agency is local (L), national (N), and whether the agency rents pickup trucks (T). Note that an agency could provide more than one service. Some of the agencies are to be randomly selected and asked various questions regarding the service they provide.

00	ADA Auto Rentals	L	10	General Motors Rental System	L,N,T
01	Alamo Rent a Car	L,N	11	Hertz Rent a Car	L,N,T
02	Avis Rent a Car	L,N	12	Lee's Sales and Service	L,T
03	Bill's Towing	L,T	13	National Car Rental	L,N
04	Budget Car and Truck Rental	L,N,T	14	Quality Imports	L
05	Charlie's Dodge	L,T	15	Rent a Wreck	L,T
06	EZ Rent a Car	L	16	Sears Car and Truck Rental	L,N,T
07	Enterprise	L,N	17	Thrifty Car Rental	L,N
08	Ford Rental System	L,N,T	18	Toyota Car Rental	L,N,T
09	Guardian Car Rentals	L,N	19	Wagoner Motor Sales	L,T

Use Appendix E – Table of Random Numbers for the following problems:

a. Randomly select a sample of five rental agencies. Start with row two and column two. The number is 90935. Use the first two numbers starting with 90.

a.

b. Randomly select a second sample of five rental agencies starting at the bottom left corner of the table. The number is 11084. Use the far right two digits starting with 84.

b.

c. Randomly select a sample of six that consists of every third rental agency starting with 02.

c.

d. Select a sample of four rental agencies so that one of each type is included. Start at the top of column three. The number is 75997. Use the left two digits starting with 75.

d.

e. Randomly select a sample of five rental agencies that rent trucks. Start at the bottom of the third column of the table. The number is 78957. Use the far right two digits starting with 57.

e.

f. Randomly select a sample of two rental agencies that rent locally and nationally, but **do not** rent trucks. Start at the bottom of the fourth column of the table. The number is 77353. Use the second and third set of digits starting with 73.

f.

g. Use the table of random numbers to select your sample of five rental agencies. Specify in your answer where you started in the random number table and how you chose that starting point.

g.

12. Five bundles of pencils contain the quantities shown at the right.

Bundle	Number of pencils
1	10
2	6
3	10
4	11
5	12

a. How many different samples of 2 bundles each are there?

a.

b. List all possible samples of size 2 and compute the mean of each sample.

Sample Number	Bundle	Total Pencils	Mean Number of Pencils

c. Calculate the population mean and compare it to the mean of the sampling distribution.

c.

13. The quality assurance department for Pepsi Distributors, Inc. maintains meticulous records on the bottling line for two-liter Pepsi bottles. Records indicate that the process follows the normal probability distribution with a mean amount per bottle of 2.01 liters and a standard deviation of 0.025 liters. The foreman randomly selects 25 bottles from the bottling line and determines that the mean amount per bottle is 2.005 liters.

a. Compute the sampling error.

a.

b. Compute the standard error of the sampling distribution of sample means.

b.

c. Compute the probability that the sample of 25 bottles would have a mean of 2.005 liters or more.

c.

14. Suppose the foreman in Problem 13 selects a second sample of 16 bottles and determines that the mean is 1.994 liters. Compute the probability that the sample of 16 bottles would have a mean of 1.994 liters or more.

14.

15. Captain D's tuna is sold in cans that have a net weight of 8 ounces. The weights are normally distributed with a mean of 8.025 ounces and a standard deviation of 0.125 ounces. You take a sample of 36 cans. Compute the probability that the sample would have a mean:

a. greater than 8.03 ounces?

a.

b. less than 7.995 ounces?

b.

c. between 7.995 and 8.03 ounces?

c.

16. The mean hourly wage of finish carpenters in Phoenix, AZ is $16.50 per hour. What is the likelihood that we could select a sample of 40 carpenters with a mean of $16.75 or more? The standard deviation of the sample is $1.75.

16.

17. The anticipated mean weight of logging trucks entering the Hafner Saw mill in southern Ohio is 78,000 pounds. A sample of 48 trucks has a mean of 78,700 pounds and a standard deviation of 1,750 pounds. What is the probability that a sample of this size could come from the population with a mean of 78,000 pounds?

17.

CHAPTER 9
ESTIMATION AND CONFIDENCE INTERVALS

Chapter Goals

After completing this chapter, you will be able to:

1. Define a ***point estimate***.
2. Define ***level of confidence***
3. Construct a confidence interval for the population mean when the population standard deviation is known.
4. Construct a confidence interval for a population mean when the population standard deviation is unknown.
5. Construct a confidence interval for a population proportion.
6. Determine the sample size for attribute and variable sampling.

Introduction

The previous chapter introduces sampling as a way to find information about a population. We noted that it is usually not possible or necessary to inspect the entire population. We gave the following as reasons for sampling:

1. *To contact the whole population would often be very time consuming*. To ask every eligible voter if they plan to vote for the current senator in the forthcoming election would take months. The election would probably be over before the survey was completed.
2. *The cost of studying all the items in the population is often prohibitive*. Some television program ratings are established by analyzing the viewing habits of about 1,200 viewers. The cost of studying all the homes having television would be exorbitant.
3. *The physical impossibility of checking all the items in the population*. The South Dakota Game Commission, for example, cannot check all the deer, grouse, and other wild game because they are always moving.
4. *The destructive nature of certain tests*. The manufacturer of fuses cannot test all of them because in the testing the fuse is destroyed and none would be available for sale.
5. *The adequacy of sample results*. If the sample results of the viewing habits of 1,200 homes revealed that only 1.1 percent of the homes watched "60 Minutes," no doubt the program would be replaced by another show. Checking the viewing habits of all the homes regarding "60 Minutes" probably would not change the percent significantly.

We also made assumptions about the population, such as the mean, the standard deviation, or the shape of the distribution of the population. We note here that in most business situations such information is not known and the purpose of sampling may be to estimate some of these values.

This chapter considers several important aspects of sampling, such as point estimates and confidence intervals.

Point Estimates and Confidence Intervals

Known σ or a Large Sample

In many situations the population is large or it is difficult to identify all the members, so we need to rely on sample information. A single number used to estimate a population parameter is called a ***point estimate***.

> ***Point estimate***: The statistic, computed from sample information, which is used to estimate the population parameter.

- The sample mean, $\overline{X}$, is a point estimate of the population mean, μ.
- The sample proportion, p, is a point estimate of the population proportion, π.
- The sample standard deviation, s, is a point estimate of the population standard deviation, σ.

For example, a sample of 100 bank tellers reveals a mean starting salary of $30,000. The $30,000 is a point estimate. The sample mean is a point estimate of the mean starting salary of all (population) accounting graduates.

We expect the point estimate to be close to the population parameter, but we would like to measure how close it really is. We need a measure that gives us a range of values into which our point estimate will fit. We use a confidence interval for this purpose.

Confidence Interval

The range of values, within which a population parameter is expected to lie, is usually referred to as the ***confidence interval.***

> ***Confidence Interval***: A range of values constructed from sample data so the parameter occurs within that range at a specified probability. This specified probability is called the *level of confidence.*

> ***Level of confidence***: The measure of the confidence we have that an interval estimate will include the population parameter.

The ***95 percent confidence interval*** means that ninety-five percent of the sample means selected from a population will be within 1.96 standard deviations of the population mean μ.

The ***99 percent confidence interval*** means that ninety-nine percent of the sample means selected from a population will be within 2.58 standard deviations of the population mean μ.

The central limit theorem allows us to state or specify a range of values within which a population parameter, such as the population mean, can be expected to occur.

When the sample size (n) is at least 30, it is generally accepted that the central limit theorem will ensure a normal distribution of the sample means. This important consideration allows us to use the standard normal distribution, that is, z in our calculation of the confidence interval. In general the confidence interval for the mean of a sample is computed by text formula [9-1].

Confidence Interval for the Population Mean $(n \geq 30)$	$\overline{X} \pm z\frac{s}{\sqrt{n}}$	[9 – 1]

Where:

$\overline{X}$ is the sample mean.
z depends on the level of confidence.
s is the sample standard deviation.
n is the size of the sample.

Unknown Population Standard Deviation and a Small Sample

In the previous section we used the standard normal distribution to express the level of confidence. We noted that:

1. The population followed the normal distribution and the population standard deviation was known, or
2. The shape of the population was not known, but the number of observations in the sample was at least 30.

What do we do if the sample size is **not** at least 30 and we **do not** know the population standard deviation? If we can reason that the population is normal or reasonably close to normal, we can replace the normal distribution with the ***t* distribution**. Also called ***Student's t distribution***

The *t* distribution

The *t* distribution is based on the assumption that the population of interest is normal, or nearly normal.

1. It is, like the *z* distribution, a continuous distribution.
2. It is, like the *z* distribution, bell-shaped and symmetrical.
3. There is not one *t* distribution but rather a "family" of *t* distributions. All *t* distributions have a mean of 0, but their standard deviations differ according to the sample size *n*. There is a *t* distribution for a sample size of 20, another for a sample size of 22, and so on. The standard deviation for a *t* distribution with 5 observations is larger than for a *t* distribution with 20 observations.
4. The *t* distribution is more spread out and flatter at the center than is the standard normal distribution. As the sample size increases, however, the *t* distribution approaches the standard normal distribution, because the "error" in using *s* to estimate σ decreases with larger samples.

The chart on the right shows the relationship between the *t* distribution and the *z* distribution.

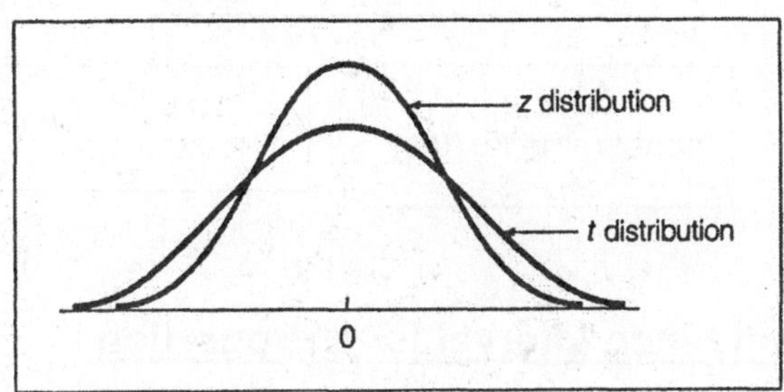

Note that the Student's t distribution has a greater spread than the z distribution, thus the value of t for a given level of confidence is larger in magnitude than the corresponding z values. Also note that for the same level of confidence the t distribution is more spread out than the z distribution.

To develop a confidence interval for the population mean with an unknown population standard deviation:

1. Assume the samples are from a normal population.
2. Estimate the population standard deviation σ with the sample standard deviation s.
3. Use the t distribution rather than the z distribution.

In order to develop a confidence interval for the population mean, using the t distribution, we adjust formula [9-1] to create formula [9-2]:

Confidence Interval for the Population Mean, σ unknown	$\bar{X} \pm t\frac{s}{\sqrt{n}}$	[9-2]

Where:
$\bar{X}$ is the sample mean.
t is the value associated with the given level of confidence.
s is the sample standard deviation.
n is the size of the sample.

The chart below summarizes the decision-making process for determining when to use the z distribution or the t distribution.

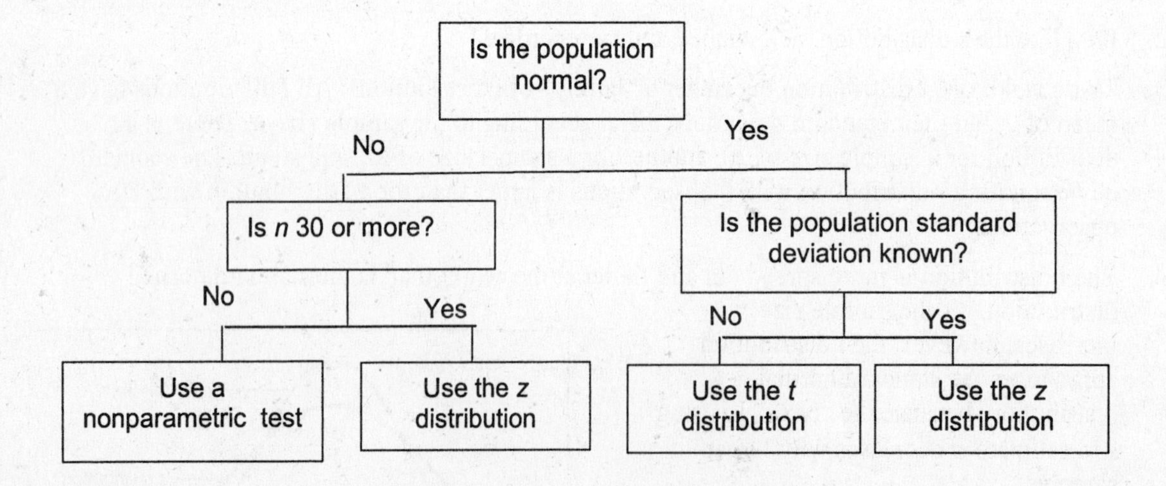

Confidence Interval for a Proportion

The previous material in this chapter dealt with the ratio scale of measurement. Variables might have been weights, lengths, distance, and income. We want to consider situations that involve the nominal scale of measurement. Recall that in the nominal scale of measurement, observations are classified into one or more mutually exclusive groups. For example, a survey may ask you if you watch "60 Minutes" on TV. Are you male or female? As a voter, are you Republican, Democrat, Independent, or other?

We are interested in knowing what ***proportion*** of a sample or population has a particular trait.

> ***Proportion*:** The fraction, ratio, or percent indicating the part of the sample or the population having a particular trait of interest.

A sample proportion can be determined using text formula [9-3]

Sample Proportion	$p = \frac{X}{n}$	[9-3]

Where:
p is the sample proportion
X is the number of successes in the sample.
n is the number of items sampled.

The population proportion is identified by π. Thus π refers to the percent of successes in the population. In Chapter 6, π is the proportion of successes in a binomial distribution.

To develop a confidence interval for a proportion we need to meet the following assumptions:

1. The binomial conditions, discussed in Chapter 6, have been met. Briefly, these conditions are:
   - **a.** The sample data is the result of counts.
   - **b.** There are only two possible outcomes. We usually label one of the outcomes a "success" and the other a "failure."
   - **c.** The probability of a success remains the same from one trial to the next.
   - **d.** The trials are independent. This means the outcome of one trial does not affect the outcome of another.
2. The values $n\pi$ and $n(1 - \pi)$ should both be greater than or equal to five. This condition allows us to invoke the central limit theorem and employ the standard normal distribution, that is z, to complete a confidence interval

Developing a point estimate and a confidence interval for a population proportion is similar to what we did for the mean. We change formula [9-1] as shown to get formula [9-4]:

Confidence Interval for a Population Proportion	$p \pm z\,\sigma_p$	[9 – 4]

Where:
p is the sample proportion
σ_p is the "standard error" of the proportion.

The "standard error" of the proportion measures the variability in the sampling distribution of the sample proportion. It is calculated using text formula [9-5].

Standard Error of the Sample Proportion	$\sigma_p = \sqrt{\frac{p(1-p)}{n}}$	[9-5]

The confidence interval for a population proportion is found by text formula [9-6]

Confidence Interval for a Population Proportion	$p \pm z\sqrt{\frac{p(1-p)}{n}}$	[9-6]

Where:
p is the sample proportion.
n is the sample size.
z is the z value for degree of confidence selected.

Finite-Population Correction Factor

If the sampling is done without replacement from a small population, the ***finite population correction factor*** is used. If the sample constitutes more than 5 percent of the population, the finite correction factor is applied. Its purpose is to account for the fact that a parameter can be more accurately estimated from a small population when a large portion of that population's units is sampled. The correction factor is:

$$\sqrt{\frac{N-n}{N-1}}$$

What is the effect of this term? If N, the number of units in the population, is large relative to n, the sample size, the value of this correction factor is near 1.00.

For example, if $N = 10{,}000$ and a sample of 40 is selected, the value of the correction factor is 0.9980, found by $\sqrt{\frac{10{,}000-40}{10{,}000-1}} = 0.9980$.

However, if N is only 500, the correction factor is 0.9601, found by $\sqrt{\frac{500-40}{500-1}} = 0.9601$.

Logically, we can estimate a population parameter with a sample of 40 from a population of 500 more accurately than with a sample of 40 from a population of 10,000.

The standard error of the mean or the standard error of the proportion is multiplied by the correction factor. Because the correction factor will always be less than 1.00, the effect is to reduce the standard error. Stated differently, because the sample constituted a substantial proportion of the population, the standard error of the sample distribution is reduced.

For a finite population, where the total number of objects is N and the size of the sample is n, the following adjustment is made to the standard error of sample means: It is formula [9-7] in the text.

Standard Error of the Sample Mean, Using a Finite Population Correction Factor	$\sigma_{\bar{X}} = \frac{\sigma}{\sqrt{n}}\sqrt{\frac{N-n}{N-1}}$	[9-7]

The confidence interval for the population mean, therefore, is computed as follows:

$$\bar{X} \pm z\frac{s}{\sqrt{n}}\left(\sqrt{\frac{N-n}{N-1}}\right)$$

For a finite population, where the total number of objects is N and the size of the sample is n, the following adjustment is made to the standard error of the proportions: It is formula [9-8] in the text.

Standard Error of the Sample Proportion, Using a Finite Population Correction Factor	$\sigma_p = \sqrt{\frac{p(1-p)}{n}}\left(\sqrt{\frac{N-n}{N-1}}\right)$	[9-8]

The confidence interval for a population proportion is:

$$p \pm z\sqrt{\frac{p(1-p)}{n}}\left(\sqrt{\frac{N-n}{N-1}}\right)$$

Choosing an Appropriate Sample Size

Sample size is always a concern when designing a statistical study. Too large a sample could be a waste of time and money collecting the data. Also, too small a sample may make the conclusions drawn from the data uncertain. The size of a sample required for a particular study is based on three factors.

1. The desired level of confidence. This is expressed in terms of z.
2. The maximum allowable error, E, the researcher will tolerate.
3. The variability in the population under study (as measured by s).

The sample size is computed using text formula [9-9]:

Sample Size for Estimating the Population Mean $n = \left(\frac{zs}{E}\right)^2$ [9 – 9]

Where:
n is the size of the sample.
z is the standard normal value corresponding to the desired level of confidence.
s is the estimate of the population standard deviation.
E is the maximum allowable error.

A population with considerable variability (reflected by a large sample standard deviation, s) will require a larger sample than a population with a smaller standard deviation. E is the maximum allowable error that you, the researcher, are willing to accept. It is the amount that is added and subtracted from the sample mean to obtain the end points of the confidence limits.

To determine the required sample size for a proportion, three items need to be specified:

1. The desired level of confidence, usually 95 percent or 99 percent.
2. The margin of error in the population proportion that is required.
3. An estimate of the population proportion π.

Text formula [9-10] is used:

Sample Size for the Population Proportion $n = p(1-p)\left(\frac{z}{E}\right)^2$ [9 – 10]

Where:
p is the estimated proportion based on the pilot survey.
z is the z score associated with the degree of confidence selected.
E is the allowable error.

If no estimate of p is available, then let $p = 0.50$. The sample size will never be larger than that obtained when $p = 0.50$.

Glossary

Confidence Interval: A range of values constructed from sample data so the parameter occurs within that range at a specified probability. This specified probability is called the *level of confidence.*

Level of confidence: The measure of the confidence we have that an interval estimate will include the population parameter.

Point estimate: The statistic, computed from sample information, which is used to estimate the population parameter.

Proportion: The fraction, ratio, or percent indicating the part of the sample or the population having a particular trait of interest.

Chapter Examples

Example 1 - Develop the 95 percent confidence interval for the population mean.

Crossett Truck Rental has a large fleet of rental trucks. At times many of the trucks need substantial repairs. Mr. Crossett has requested a study of the repair costs. A random sample of 64 trucks is selected. The mean annual repair cost is $1,200, with a standard deviation of $280. Estimate the mean annual repair cost for all rental trucks. Develop the 95 percent confidence interval for the population mean.

Solution 1 - Develop the 95 percent confidence interval for the population mean.

The population parameter being estimated is the population mean — the mean annual repair cost of all Crossett rental trucks. This value is not known, but the best estimate we have of that value is the sample mean of $1,200. Hence, $1,200 is a point estimate of the unknown population parameter. A confidence interval is a range of values within which the population parameter is expected to occur. The 95 percent refers to the approximate percent of time that similarly constructed intervals would include the parameter being estimated.

The confidence interval for a mean is obtained by applying formula [9-1].

$$\bar{X} \pm z\frac{s}{\sqrt{n}} \qquad [9-1]$$

How is the z value determined? In this Example the 95 percent level of confidence is used. This refers to the middle 95 percent of the values. The remaining 5 percent is divided equally between the two tails of curve. (See the following diagram.)

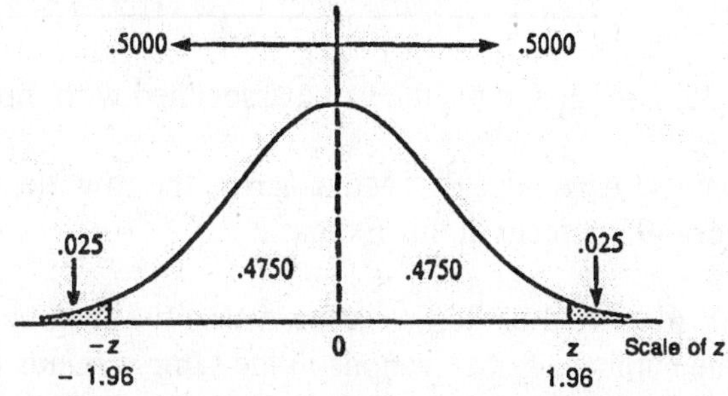

To find z, the standard normal distribution is used. Referring to Appendix D, the first step is to locate the value of 0.4750 in the body of the table, and then read the corresponding row and column values. The z value is 1.96.

Substitute the z value into the equation. The confidence interval is $1,131.40 to $1,268.60

$$\begin{aligned}\bar{X} \pm z\frac{s}{\sqrt{n}} &= \$1{,}200 \pm 1.96\frac{\$280}{\sqrt{64}} \\ &= \$1{,}200 \pm \$68.6 \\ &= \$1{,}131.40 \quad \text{to} \quad \$1{,}268.60\end{aligned}$$

This indicates that if 100 similar intervals were constructed, about 95 intervals would be expected to include the population mean.

Self-Review 9.1

Check your answers against those in the ANSWER section.

The Internal Revenue Service is studying contributions to charity. A random sample of 36 returns is selected. The mean contribution is $150 and the standard deviation of the sample is $20. Construct a 98 percent confidence interval for the population mean.

Example 2 – Constructing a 95% Confidence Interval for the Population Mean

A manufacturer of batteries for portable hand tools wishes to investigate the length of time a battery will last between charges at a fixed load. A sample of 12 batteries had a mean length of time of 4.3 hours with a standard deviation of 0.25 hours.

a. Construct a 95 percent confidence interval for the population mean.

b. Is it reasonable for the manufacturer to claim that the batteries will last 4.25 hours?

c. How about 5 hours?

Solution 2 – Constructing a 95% Confidence Interval for the Population Mean

a. Use the t distribution because the population standard deviation is unknown and the sample contains less than 30 values.

We use formula [9-2] to develop the confidence interval.

$$\overline{X} \pm t\frac{s}{\sqrt{n}} \qquad [9-2]$$

Where: $\overline{X} = 4.3$, $s = 0.25$, $n = 12$, and t is the value associated with the given level of confidence.

To find the value of t we use Appendix F. Move across the row identified as "Confidence Interval" to the level of confidence—95 percent in this Example.

The column on the left, identified as "d.f.," is known as the Degrees of Freedom. The number of degrees of freedom is the number of observations in the sample minus the number of samples, written $(n-1)$. In this case it is $(12-1) = 11$. The value of t is 2.201. We substitute these values in formula [9-2].

$$\overline{X} \pm t\frac{s}{\sqrt{n}} = 4.3 \pm 2.201\frac{0.25}{\sqrt{12}} = 4.3 \pm 0.159$$

The end points of the interval are: $4.3 - 0.159 = 4.141$ and $4.3 + 0.159 = 4.459$. Because 4.25 hours is in the interval between 4.141 hours and 4.459 hours, we conclude the population mean could be 4.25 hours.

b. The manufacturer can be 95 percent confident that the mean life between charges is 4.25 hours, since this value is in the interval.

c. Because 5.00 hours is not in the interval, we conclude a value of 5 for the population mean is not reasonable.

Self-Review 9.2

Check your answers against those in the ANSWER section.

A manufacturer of batteries for "kids' toys" wishes to investigate the length of time a battery will last. Tests results on a sample of 10 batteries indicated a sample mean of 5.67 and a sample standard deviation of 0.57.

a. Determine the mean and the standard deviation
b. What is the population mean? What is the best estimate of that value?
c. Construct a 95 percent confidence interval for the population mean.
d. Explain why the t distribution is used as a part of the confidence interval.
e. Is it reasonable for the manufacturer to claim that the batteries will last 6.0 hours?

Example 3 – Develop a 99% Confidence Interval for the Population Proportion

A market survey is conducted by your state legislator to determine the proportion of homeowners who would switch to a new "electrical energy supplier" if they had the opportunity afforded them by new state legislation. Of the 1,200 homeowners surveyed, 800 said they would switch.

a. Estimate the value of the population proportion.
b. Compute the standard error of the proportion
c. Develop a 99 percent confidence interval for the population proportion.
d. Interpret the results if the legislator states that 2/3 of the homeowners would switch.

Solution 3 – Develop a 99% Confidence Interval for the Population Proportion

a. Use formula [9-3] to estimate the value of the population proportion. $p = \frac{X}{n} = \frac{800}{1200} = 0.667$

b. Use formula [9-5] to estimate the standard error of the proportion.

$$\sigma_p = \sqrt{\frac{p(1-p)}{n}} = \sqrt{\frac{0.667(1-0.667)}{1200}} = 0.0136$$

c. The 99 percent confidence interval is found by using formula [9-6].

$$\begin{aligned} & p \pm z\sqrt{\frac{p(1-p)}{n}} \\ &= 0.667 \pm 2.58\sqrt{\frac{0.667(1-0.667)}{1200}} \\ &= 0.667 \pm 2.58 \times 0.0136 \\ &= 0.667 \pm 0.035 \\ &= 0.632 \text{ and } 0.702 \end{aligned}$$

d. The value 0.667 is in the interval, thus the legislator is correct in stating that 2/3 of the homeowners would switch energy suppliers.

Example 4 – Develop a 95% Confidence Interval for the Population Mean

Refer to the information on Crossett Truck Rental in Example 1. Suppose Crossett's fleet consists of 500 trucks. Develop a 95 percent confidence interval for the population mean.

Solution 4 – Develop a 95% Confidence Interval for the Population Mean

When the sample is more than 5 percent of the population, the finite population correction factor is used. In this case the sample size is 64 and the population size is 500. Thus, $n/N = 64/500 = 0.128$ or 12.8 percent. The confidence interval is adjusted as follows:

$$\overline{X} \pm z\frac{s}{\sqrt{n}}\left(\sqrt{\frac{N-n}{N-1}}\right) = \$1{,}200 \pm 1.96\frac{\$280}{\sqrt{64}}\left(\sqrt{\frac{500-64}{500-1}}\right)$$

$$= \$1{,}200 \pm \$68.6(0.9347)$$

$$= \$1{,}200 \pm \$64.12$$

$$= \$1{,}135.88 \quad \text{to} \quad \$1{,}264.12$$

Notice that when the correction factor is included, the confidence interval becomes smaller. This is logical because the number of items sampled is large relative to the population.

Self-Review 9.3

Check your answers against those in the ANSWER section.

Refer to Self-Review 9.1. Compute the 98 percent confidence interval if the population consists of 200 tax returns.

Example 5 – Develop a 99% Confidence Interval for the Population Proportion

The Independent Department Store wants to determine the proportion of their charge accounts that have an unpaid balance of $1,500 or more. A sample of 250 accounts revealed that 100 of them had an unpaid balance of $1,500 or more. What is the 99 percent confidence interval for the population proportion? Would it be reasonable to conclude that more than half of the account balances are more than $1500?

Solution 5 – Develop a 99% Confidence Interval for the Population Proportion

In the sample of 250 charge accounts, there were 100 with unpaid balances of over $1,500. The point estimate of the proportion of charge customers with balances of more than $1,500 is 0.40, found by 100 ÷ 250. The z value corresponding to a 99 percent level of confidence is 2.58 (from Appendix D). The formula for the confidence interval for the population proportion is text formula [9-6]:

$$p \pm z\sqrt{\frac{p(1-p)}{n}} = 0.40 \pm 2.58\sqrt{\frac{(0.40)(1-0.40)}{250}} = 0.40 \pm (2.58)\sqrt{0.00096} = 0.40 \pm 0.08$$

The confidence interval is 0.32 to 0.48. This means that about 99 percent of the similarly constructed intervals would include the population proportion. Because 0.50 is not in the interval, we cannot conclude that more than half the account balances are over $1500.

Self-Review 9.4

Check your answers against those in the ANSWER section.

A random sample of 100 light bulbs is selected. Sixty were found to burn for more than 1,000 hours. Develop a 90 percent confidence interval for the proportion of bulbs that will burn more than 1,000 hours.

Example 6 - Develop a 99 percent confidence interval

Refer to the charge account data of Independent Department Stores in Example 5. Recall that 250 accounts were sampled. Suppose there is a total of 900 charge customers. Develop a 99 percent confidence interval for the proportion of charge customers with an unpaid account balance over $1,500.

Solution 6 - Develop a 99 percent confidence interval

The finite population correction factor should be used because the sample is 28 percent of the population, found by 250 ÷ 900.

$$p \pm z\sqrt{\frac{p(1-p)}{n}}\left(\sqrt{\frac{N-n}{N-1}}\right) = 0.40 \pm 2.58\sqrt{\frac{(0.40)(1-0.40)}{250}}\left(\sqrt{\frac{900-250}{900-1}}\right)$$

$$= 0.40 \pm 0.08(0.8503)$$

$$= 0.40 \pm 0.068$$

Using the correction factor, the interval is reduced from 0.40 ± 0.08 to 0.40 ± 0.068, or 0.332 to 0.468. Again, this is because Independent Stores has sampled a large proportion (28 percent) of its customers.

Example 7 – Determining the Size of the Sample

The manager of the Jiffy Supermarket wants to estimate the mean time a customer spends in the store. Use a 95 percent level of confidence. The standard deviation of the population based on a pilot survey is estimated to be 3.0 minutes. The manager requires the estimate to be within plus or minus 1.00 minute of the population value. What sample size is needed?

Solution 7 – Determining the Size of the Sample

The size of the sample is dependent on three factors.

1. The allowable error (E).
2. The level of confidence (z).
3. The estimated variation in the population, usually measured by s, the sample standard deviation.

In this Example, the store manager has indicated that the estimate must be within 1.0 minute of the population parameter. The level of confidence is 0.95 and the population standard deviation is estimated to be 3.0 minutes. The formula [9-9] for determining the size of the sample is:

$$n = \left(\frac{zs}{E}\right)^2 = \left(\frac{(1.96)(3.0)}{1.0}\right)^2 = (5.88)^2 = 34.57 = 35$$

Hence, the manager should randomly select 35 customers and determine the amount of time they spend in the store.

Self-Review 9.5

Check your answers against those in the ANSWER section.

A health maintenance organization (HMO) wants to estimate the mean length of a hospital stay. How large a sample of patient records is necessary if the HMO wants to be 99 percent confident of the estimate and wants the estimate to be within plus or minus 0.2 days? An earlier study showed the standard deviation of the length of stay to be 0.25 days.

Example 8 – Determining the Size of the Sample

The Ohio Unemployment Commission wants to estimate the proportion of the labor force that was unemployed during last year in a certain depressed region. The Commission wants to be 95 percent confident that their estimate is within 5 percentage points (written 0.05) of the population proportion. If the population proportion has been estimated to be 0.15, how large a sample is required?

Solution 8 – Determining the Size of the Sample

Note that the estimate of the population proportion is 0.15. The allowable error (E) is 0.05. Using the 95 percent level of confidence, the z value is 1.96. Applying formula [9-10] to determine the sample size:

$$n = p(1-p)\left(\frac{z}{E}\right)^2 = 0.15(1-0.15)\left(\frac{1.96}{0.05}\right)^2 = 195.92 = 196$$

The required sample size is 196. When no estimate of population proportion is available, 0.50 is used. The size of the sample will never be larger than that obtained when $p = 0.50$. The calculations for the sample size when $p = 0.50$ are:

$$n = p(1-p)\left(\frac{z}{E}\right)^2 = 0.5(1-0.5)\left(\frac{1.96}{0.05}\right)^2 = 384.16 = 385$$

Note that the required sample size is considerably larger (385 versus 196) when p is set at 0.50.

Self-Review 9.6

Check your answers against those in the ANSWER section.

A large bank believes that one-third of its checking customers have used at least one of the bank's other services during the past year. How large a sample is required to estimate the actual proportion within a range of plus and minus 0.04? Use the 98 percent level of confidence.

CHAPTER 9 ASSIGNMENT

ESTIMATION AND CONFIDENCE INTERVALS

Name ______________________ Section __________ Score ________

Part I Select the correct answer and write the appropriate letter in the space provided.

______ 1. A single number used to estimate a population parameter is
a. the confidence interval.
b. the population parameter.
c. a point estimate.
d. the mean of the population.

______ 2. A range of values constructed from sample data so that the parameter occurs within that range at a specified probability is
a. a confidence interval.
b. the population parameter.
c. a point estimate.
d. the mean of the population

______ 3. The size of the standard error is affected by the standard deviation of the sample and
a. a confidence interval.
b. the population parameter.
c. the point estimate.
d. the sample size

______ 4. Suppose we select 100 samples from a population. For each sample we construct a 95 percent confidence interval. We could expect about 95 percent of these confidence intervals to contain
a. a sample mean.
b. the population mean.
c. a point estimate.
d. the standard deviation of the population

______ 5. The *t* distribution is a continuous distribution, with many similarities to
a. the confidence interval.
b. the population parameter.
c. the standard normal distribution.
d. the mean of the population

______ 6. The *t* distribution is used when the population is normal, the sample is less than 30, and
a. the population standard deviation is unknown.
b. the population standard deviation is known.
c. the point estimate is known.
d. the mean of the population is unknown.

______ 7. If the level of confidence is decreased from 95 percent to 90 percent, the width of the corresponding interval will
a. be increased.
b. be decreased.
c. stay the same.
d. not have an effect on the level of confidence

______ 8. The finite population correction factor is used when
a. the sample is more than 5 percent of the population.
b. the sample is less than 5 percent of the population.
c. the sample is larger than the population.
d. the population cannot be estimated.

_____ 9. A 90 percent confidence interval for means indicates that 90 out of 100 similarly constructed intervals will include the

a. sample mean. b. sampling error.
c. z value d. population mean.

_____ 10. The fraction, ratio, or percent indicating the part of the sample or the population having a particular trait of interest is

a. a confidence interval. b. the population parameter.
c. a point estimate. d. the proportion.

Part II Answer the following questions. Be sure to show essential work.

11. As part of a safety check, the Pennsylvania Highway Patrol randomly stopped 65 cars and checked their tire pressure. The sample mean was 32 pounds per square inch with a sample standard deviation of 2 pounds per square inch. Develop a 98 percent confidence interval for the population mean.

11. _____ to _____

12. A survey of 4,000 college graduates determines that the mean length of time to earn a bachelor's degree is 5.08 years and the standard deviation is 1.89 years. Construct a 96 percent confidence interval for the mean time required for all graduates to earn a bachelor's degree.

12. _____ to _____

13. Suppose the college in question 12 has only graduated 10,000 students. Construct a 96 percent confidence interval for the mean time required for all graduates to earn a bachelor's degree.

13. _____ to _____

14. A manufacturer of diamond drill bits for industrial production drilling and machining wishes to investigate the length of time a drill bit will last while drilling carbon steel. The production of the drill bits is very expensive, thus the number available for testing is small. A sample of 8 drill bits had a mean drilling time of 2.25 hours with a standard deviation of 0.5 hours.

a. Construct a 95 percent confidence interval for the population mean.

14 a. to

b. Is it reasonable for the manufacturer to claim that the drill bits will last 2.5 hours?

____

15. Of a random sample of 90 firms with employee stock ownership plans, 50 indicated that the primary reason for setting up the plan was tax related. Develop a 90 percent confidence interval for the population proportion of all such firms with this as the primary motivation.

15. to

16. A study of 305 computer chips found that 244 chips functioned properly. Develop a 99 percent confidence interval for the population proportion of properly functioning computer chips.

16. to

17. A correctional institution would like to report the mean amount of money spent per day on operating the facilities. How many days should be considered if a 95 percent confidence is used and the estimate is to be within one hundred dollars? The standard deviation is $400.

17.

18. ***The Corporate Lawyer***, a magazine for corporate lawyers, would like to report the mean amount earned by lawyers in their area of specialization. How large a sample is required if the 97 percent level of confidence is used and the estimate is to be within $2,500? The standard deviation is $16,000.

18.

19. The Customer Relations Department at Commuter Airline, Inc. wants to estimate the proportion of customers that carry only hand luggage. The estimate is to be within 0.03 of the true proportion with 95 percent level of confidence. No estimate of the population proportion is available. How large a sample is required?

19.

20. A survey is being conducted on a local mayoral election. If the poll is to have a 98 percent confidence interval and must be within four percentage points, how many people should be surveyed?

20.

CHAPTER 10
ONE-SAMPLE TESTS OF HYPOTHESIS

Chapter Goals

After completing this chapter, you will be able to:

1. Define a hypothesis and hypothesis testing.
2. Describe the five-step hypothesis-testing procedure.
3. Distinguish between a one-tailed and a two-tailed test of hypothesis.
4. Conduct a test of hypothesis about a population mean.
5. Conduct a test of hypothesis about a population proportion.
6. Define *Type I* and *Type II* errors.

Introduction

In Chapter 8 we began our study of statistical inference by describing how we could select a random sample and then use the sample values to estimate the value of a population parameter. Recall that a sample is a part or subset of the population, while a parameter is a value calculated from the entire population. In Chapter 9 we estimated a population parameter from a sample statistic. In addition, we developed a range of values, called a confidence interval, within which we expected the population value to be located.

In this chapter, rather than developing a range of values within which we expect the population parameter to occur, we will conduct a test of hypothesis regarding the validity of a statement about a population parameter.

Two statements called hypotheses are made regarding the possible values of population parameters.

What is a Hypothesis?

A ***hypothesis*** is a statement about a population.

> ***Hypothesis***: A statement about a population parameter developed for the purpose of testing.

In statistical analysis we make a claim, that is, state a hypothesis, and then follow up with tests to verify the assertion or to determine that it is untrue.

What is Hypothesis Testing?

The terms ***hypothesis testing*** and ***testing a hypothesis*** are used interchangeably. Hypothesis testing starts with a statement about a population parameter such as the mean.

> ***Hypothesis testing***: A procedure based on sample evidence and probability theory to determine whether the hypothesis is a reasonable statement.

For example, one statement about the performance of a new model car is that the mean miles per gallon is 30. The other statement is that the mean miles per gallon is not 30. Only one of these statements is correct.

Five-Step Procedure for Testing a Hypothesis

Statistical hypothesis testing is a five-step procedure. These steps are:

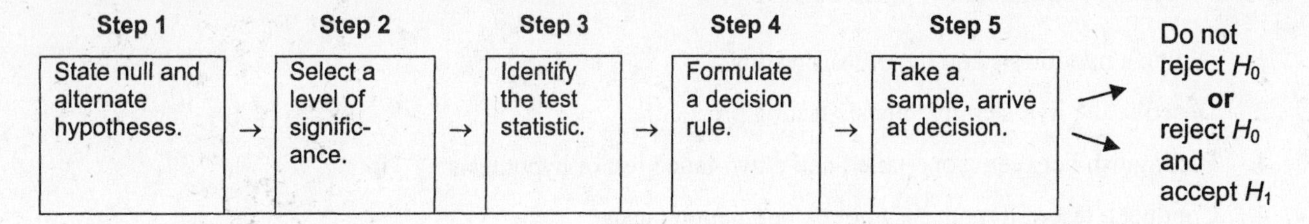

When conducting hypothesis tests, we actually employ a strategy of "proof by contradiction." That is, we hope to accept a statement to be true by rejecting or ruling out another statement. The steps involved in hypothesis testing will now be described in more detail.

First we will concentrate on testing a hypothesis about a population mean, or means. Then we will consider one or two population proportions. For a mean or means:

Step 1. State the null hypothesis (H_0) and the alternate hypothesis (H_1).

The first step is to state the hypothesis being tested. It is called the ***null hypothesis***, designated H_0, and read H sub zero. The capital letter H stands for hypothesis, and the subscript zero implies "no difference."

> ***Null hypothesis***: A statement about the value of a population parameter.

For example, a recent newspaper report made the claim that the mean length of a hospital stay was 3.3 days. You think that the true length of stay is some other length than 3.3 days.

The null hypothesis is written H_0: $\mu = 3.3$, where H_0 is an abbreviation of the null hypothesis. The null hypothesis will always contain the equal sign. It is the statement about the value of the population parameter, in this case the population mean. The null hypothesis is established for the purpose of testing. On the basis of the sample evidence, it is either rejected or not rejected.

If the null hypothesis is rejected then we accept the ***alternate hypothesis***.

> ***Alternate hypothesis***: A statement that is accepted if the sample data provide enough evidence that the null hypothesis is false.

The alternate hypothesis is written H_1. From the above example the alternate hypothesis is that the mean length of stay is not 3.3 days. It is written H_1: $\mu \neq 3.3$ ($\neq$ is read "not equal to"). H_1 is accepted only if H_0 is rejected. When the "$\neq$" sign appears in the alternate hypothesis, the test is called **a two-tailed test**.

There are two other formats for writing the null and alternate hypotheses. Suppose you think that the mean length of stay is greater than 3.3 days. The null and alternate hypotheses would be written as follows: ($\leq$ is read "equal to or less than").

$$H_0\colon \mu \leq 3.3$$
$$H_1\colon \mu > 3.3$$

Notice that in this case the null hypothesis indicates "no change or that μ is less than 3.3." The alternate hypothesis states that the mean length of stay is greater than 3.3 days. Acceptance of the alternate hypothesis would allow us to conclude that the mean length of stay is greater than 3.3 days.

What if you think that the mean length of stay is less than 3.3 days? The null and alternate hypotheses would be written as:

$$H_0\colon \mu \geq 3.3$$
$$H_1\colon \mu < 3.3$$

Acceptance of the alternate hypothesis in this instance would allow you to conclude the mean length of stay is less than 3.3 days. When a direction is expressed in the alternate hypothesis, such as $>$ or $<$, the test is referred to as being **one-tailed.**

Step 2. Select a Level of Significance.

After you establish the null hypothesis and alternate hypothesis, the next step is to state the ***level of significance.***

> ***Level of significance***: The probability of rejecting the null hypothesis when it is true.

The level of significance is designated α, the Greek letter alpha. The level of significance is sometimes called the level of risk. It will indicate when the sample mean is too far away from the hypothesized mean for the null hypothesis to be true. Usually the significance level is set at either 0.01 or 0.05, although other values may be chosen.

Testing a null hypothesis at the 0.05 significance level, for example, indicates that the probability of rejecting the null hypothesis, even though it is true, is 0.05. The 0.05 level is also stated as the 5% level. When a true null hypothesis is rejected, it is referred to as a ***Type I error.***

> ***Type I error***: Rejecting the null hypothesis, H_0, when it is true.

The decision whether to use the 0.01 or the 0.05 significance level, or some other value, depends on the consequences of making a Type I error. The significance level is chosen before the sample is selected.

If the null hypothesis is not true, but our sample results indicate that it is, we have a ***Type II error.***

> ***Type II error***: Accepting the null hypothesis when it is false.

For example, H_0 is that the mean hospital stay is 3.3 days. Our sample evidence fails to refute this hypothesis, but actually the population mean length of stay is 4.0 days. In this situation we have committed a Type II error by accepting a false H_0.

We refer to the probability of these two possible errors as *alpha* α and *beta* β. *Alpha* (α) is the probability of making a Type I error and *beta* (β) is the probability of making a Type II error. The table on the right summarizes the decisions the researcher could make and the possible consequences.

	Researcher	
Null Hypothesis	**Accepts H_0**	**Rejects H_0**
H_0 is true	Correct decision	Type I error
H_0 is false	Type II error	Correct decision

Step 3. Select the Test Statistic.

A ***test statistic*** is a quantity calculated from the sample information and is used as the basis for deciding whether or not to reject the null hypothesis.

> ***Test statistic***: A value, determined from sample information, used to determine whether to reject the null hypothesis.

Exactly which test statistic to employ is determined by factors such as whether the population standard deviation is known and the size of the sample.

In hypothesis testing for the mean μ, when σ is known or the sample size is large, the standard normal distribution, the z value, is the test statistic used. formula [10-1] is used:

z distribution as a Test Statistic	$z = \dfrac{\overline{X} - \mu}{\sigma / \sqrt{n}}$	$[10-1]$

Where:

z is the value of the test statistic.
$\overline{X}$ is the sample mean.
μ is the population mean.
σ is the population standard deviation.
n is the sample size.

Step 4. Formulate the Decision Rule.

A ***decision rule*** is based on H_0 and H_1, the level of significance, and the test statistic.

> ***Decision rule***: A statement of the conditions under which the null hypothesis is rejected and conditions under which it is not rejected.

The region or area of rejection indicates the location of the values that are so large or so small that the probability of their occurrence for a true null hypothesis is rather remote.

If we are applying a one-tailed test, there is one ***critical value***. If we are applying a two-tailed test, there are two critical values.

> ***Critical value***: The dividing point between the region where the null hypothesis is rejected and the region where it is not rejected.

Chart 10-1 shows the conditions under which the null hypothesis is rejected, using the 0.05 significance level, a one-tailed test, and the standard normal distribution.

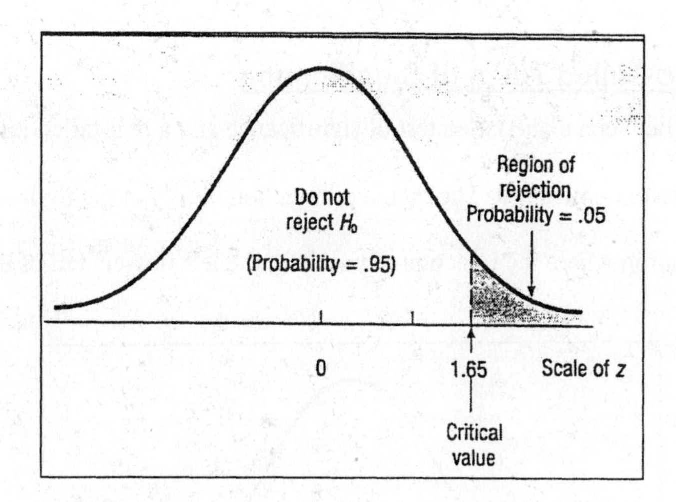

Chart 10-1 Sampling Distribution of the Statistic z, Right-Tailed Test, 0.05 Level of Significance

The above diagram portrays the rejection region for a right-tailed test.

1. The area where the null hypothesis is not rejected is to the left of 1.65.
2. The area of rejection is to the right of 1.65.
3. A one-tailed test is being applied.
4. The 0.05 level of significance was chosen.
5. The sampling distribution of the test statistic z is normally distributed.
6. The value 1.65 separates the regions where the null hypothesis is rejected and where it is not rejected.
7. The value 1.65 is called the *critical value.*

When is the standard normal distribution used? It is appropriate when the population is normal and the population standard deviation is known. When the population standard deviation is not known, the sample standard deviation is used instead. If the sample is at least 30, the test statistic follows the normal distribution.

If the computed value of z is greater than 1.65, the null hypothesis is rejected. If the computed value of z is less than or equal to 1.65, the null hypothesis is not rejected.

Step 5. Compute the value of the test statistic, make a decision, and interpret the results.

The final step in hypothesis testing after selecting the sample is to compute the value of the test statistic. This value is compared to the critical value, or values, and a decision is made whether to reject or not to reject the null hypothesis. Interpret the results.

A summary of the steps in hypothesis testing:

1. Establish the null hypothesis (H_0) and the alternate hypothesis (H_1).
2. Select the level of significance, that is α.
3. Select an appropriate test statistic.
4. Formulate the decision rule, based on steps 1, 2, and 3 above.
5. Make a decision regarding the null hypothesis based on the sample information. Interpret the results of the test.

One-Tailed and Two-Tailed Tests of Significance

We need to differentiate between a one-tailed test of significance and a two-tailed test of significance.

Chart 10-1 above depicts a one-tailed test. The region of rejection is only in the right (upper) tail of the curve.

Chart 10-2 depicts a situation where the rejection region is in the left (lower) tail of the normal distribution.

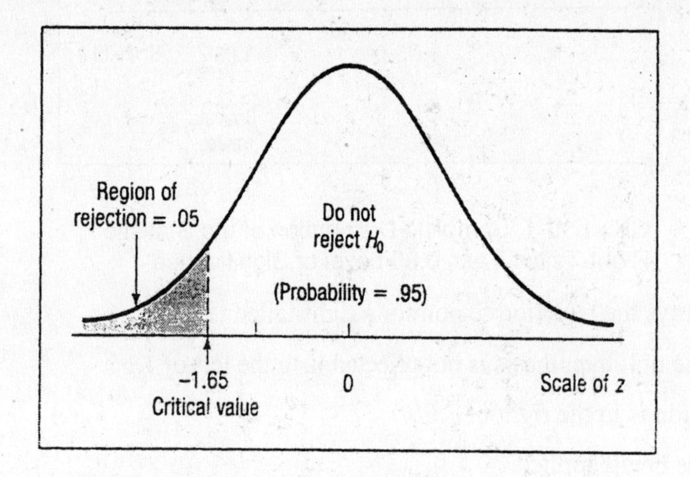

Chart 10-2 Sampling Distribution of the Statistic z, Left-Tailed Test, 0.05 Level of Significance

Chart 10-3 depicts a situation for a two-tailed test where the rejection region is divided equally into the two tails of the normal distribution.

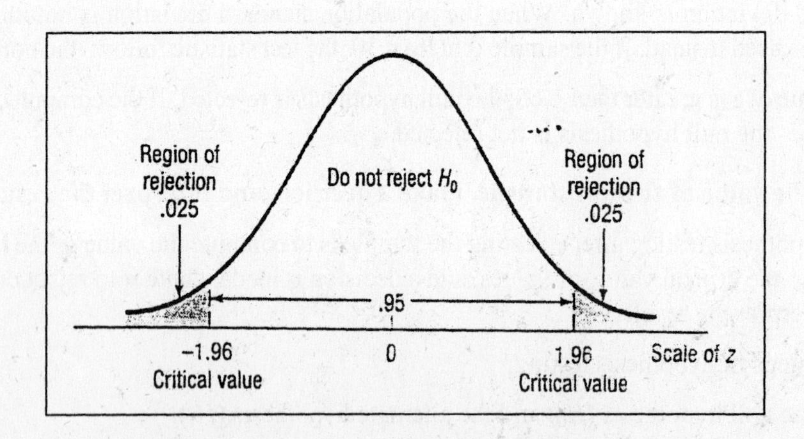

Chart 10-3 Regions of Nonrejection and Rejection for a Two-Tailed Test, 0.05 Level of Significance

Testing for a Population Mean with a Known Population Standard Deviation

Suppose we are concerned with a single population mean. We want to test if our sample mean could have been obtained from a population with a particular hypothesized mean. For example, we may be interested in testing whether the mean starting salary of recent marketing graduates is equal to $32,000 per year. It is assumed that:

1. The population is normally distributed.
2. The population standard deviation is known.

If σ is not known, the sample standard deviation is substituted for the population standard deviation provided the sample size is 30 or more.

Under these conditions the test statistic is the standard normal distribution with the sample standard deviation s substituted for σ. Thus we use text formula [10-1].

$$z = \frac{\overline{X} - \mu}{\sigma / \sqrt{n}}$$

Where:

z is the value of the test statistic.
$\overline{X}$ is the sample mean.
μ is the population mean.
σ is the standard deviation of population.
n is the number in sample.

The sample standard deviation s can be substituted for σ providing that the sample size is 30 or more.

p-value in Hypothesis Testing

In the process of testing a hypothesis, we compared the test statistic to a critical value. We made a decision to either reject the null hypothesis or not to reject it. The question is often asked as to how confident we were in rejecting the null hypothesis.

A ***p-value*** is frequently compared to the significance level to evaluate the decision regarding the null hypothesis. It is a means of reporting the likelihood that H_0 is true.

> ***p-value*:** The probability of observing a sample value as extreme as, or more extreme than, the value observed, given that the null hypothesis is true.

- If the *p-value* is greater than the significance level, then H_0 is not rejected.
- If the *p-value* is less than the significance level, then H_0 is rejected.
- The *p*-value for a given test depends on three factors:
    1. whether the alternate hypothesis is one-tailed or two-tailed
    2. the particular test statistic that is used
    3. the computed value of the test statistic

For example, if $\alpha = 0.05$ and the *p*-value is 0.0025, H_0 is rejected. We report there is only a 0.0025 likelihood that H_0 is true.

Interpreting the weight of evidence against H_0.	If the p value is less than (a) 0.10, we have *some* evidence that H_0 is not true. (b) 0.05, we have *strong* evidence that H_0 is not true. (c) 0.01, we have *very strong* evidence that H_0 is not true. (d) 0.001, we have *extremely strong* evidence that H_0 is not true.

Testing for a Population Mean: Large Sample, Population Standard Deviation Unknown

In most cases the population standard deviation is unknown. Thus, σ must be based on prior studies or estimated by the sample standard deviation, s. As long as the sample size, n, is at least 30, s can be substituted for σ, as illustrated in the following formula.

z Statistic, σ Unknown

$$z = \frac{\bar{X} - \mu}{s / \sqrt{n}} \qquad [10\text{-}2]$$

Tests Concerning Proportions

In the previous chapter we discussed confidence intervals for proportions. We continue our study of hypothesis testing but expand the idea to a ***proportion***. What is a proportion?

> ***Proportion***: The fraction, ratio, or percent indicating the part of the population or sample having a particular trait of interest.

If we let p stand for the sample proportion then text formula [10-3] is:

Test of Hypothesis, One Proportion

$$z = \frac{p - \pi}{\sigma_p} \qquad [10-3]$$

Where:

z is the value of the test statistic

π is the population proportion.

p is the sample proportion.

σ_p is the standard error of the population proportion. It is computed by $\sqrt{\pi(1-\pi)/n}$ so the formula for z becomes text formula [10–4]:

Test of Hypothesis, One Proportion

$$z = \frac{p - \pi}{\sqrt{\frac{\pi(1-\pi)}{n}}} \qquad [10-4]$$

Where:

z is the value of the test statistic

π is the population proportion.

p is the sample proportion.

n is the sample size.

For example, we want to estimate the proportion of all home sales made to first time buyers. A random sample of 200 recent transactions showed that 40 were first time buyers. Therefore, we estimate that 0.20, or 20 percent, of all sales are made to first time buyers, found by:

$$p = \frac{40}{200} = 0.20$$

To conduct a test of hypothesis for proportions, the same assumptions required for the binomial distribution must be met. Recall from Chapter 6 that those assumptions are:

1. Each outcome is classified into one of two categories such as, buyers were either first time home buyers or they were not.
2. The number of trials is fixed. In this case it is 200.
3. Each trial is independent, meaning that the outcome of one trial has no bearing on the outcome of any other. Whether the 20th sampled person was a first time buyer does not affect the outcome of any other trial.
4. The probability of a success is fixed. The probability is 0.20 for all 200 buyers in the sample.

Recall from Chapter 6 that the normal distribution is a good approximation of the binomial distribution when $n\pi$ and $n\,(1 - \pi)$ are both greater than 5. In this instance n refers to the sample size and π to the probability of a success. The test statistic that is employed for testing hypotheses about proportions is the standard normal distribution.

Testing for a Population Mean: Small Sample, Population Standard Deviation Unknown

Recall that we can use the standard normal distribution, that is z, when:

1. The population is known to follow a normal distribution and the population standard deviation is known, or
2. The shape of the population is not known, but the sample size is at least 30.

When the population standard deviation is not known and the sample size is at least 30 the correct statistical procedure is to replace the standard normal distribution with the t distribution. In Chapter 9, we noted that the following major characteristics of the t distribution:

1. It is a continuous distribution.
2. It is bell shaped and symmetrical.
3. There is a "family" of t distributions. Each time the size of the sample changes, and thus the degrees of freedom change, a new t distribution is created.
4. As the number of degrees of freedom increases, the shape of the t distribution approaches that of the standard normal distribution.
5. The t distribution is more spread out (that is, "flatter") than the standard normal distribution.

To conduct a test of hypothesis using the t distribution, we use text formula [10-5]:

t Statistic	$t = \dfrac{\overline{X} - \mu}{s / \sqrt{n}}$	[10-5]

Where:

t is the value of the test statistic.
$\overline{X}$ is the mean of the sample.
μ is the hypothesized population mean.
s is the standard deviation of the sample.
n is the number of observations in the sample.

Types of Tests of Hypothesis for a Proportion

There are three formats for testing a hypothesis about a proportion. For a one-tailed test there are two possibilities, depending on the intent of the researcher. For example, if we wanted to determine whether more than 25 percent of the sales of homes were sold to first time buyers, the hypotheses would be given as follows:

$$H_0: \pi \leq 0.25$$
$$H_1: \pi > 0.25$$

If we wanted to find out whether fewer than 25 percent of the homes were sold to first time buyers, the hypotheses would be given as:

$$H_0: \pi \geq 0.25$$
$$H_1: \pi < 0.25$$

For a two-tailed test the null and alternate hypotheses are:

$$H_0: \pi = 0.25$$
$$H_1: \pi \neq 0.25$$

Where $\neq$ means "not equal to." Rejection of H_0 and acceptance of H_1 allows us to conclude only that the population proportion is "different from" or "not equal to" the population value. It does not allow us to make any statement about the direction of the difference.

Glossary

Alternate hypothesis: A statement that is accepted if the sample data provide evidence that the null hypothesis is false.

Critical value: The dividing point between the region where the null hypothesis is rejected and the region where it is not rejected.

Decision rule: A statement of the conditions under which the null hypothesis is rejected and conditions under which it is not rejected.

Hypothesis: A statement about a population parameter developed for the purpose of testing.

Hypothesis testing: A procedure based on sample evidence and probability theory to determine whether the hypothesis is a reasonable statement.

Level of significance: The probability of rejecting the null hypothesis when it is true.

Null hypothesis: A statement about the value of a population parameter.

p-value: The probability of observing a sample value as extreme as, or more extreme than, the value observed, given that the null hypothesis is true.

Proportion: The fraction, ratio, or percent indicating the part of the population or sample having a particular trait of interest

Test statistic: A value, determined from sample information, used to determine whether to reject the null hypothesis.

Type I error: Rejecting the null hypothesis, H_0, when it is true.

Type II error: Accepting the null hypothesis when it is false.

Chapter Examples

Example 1 – Five-Step Procedure for Testing a Hypothesis

The manufacturer of the new subcompact Clipper claims in their TV advertisements that it will average "40 or more miles per gallon on the open road." Some of the competitors believe this claim is too high. To investigate, an independent testing agency is hired to conduct highway mileage tests. A random sample of 64 Clippers showed their mean miles per gallon to be 38.9, with a sample standard deviation of 4.00 miles per gallon. At the 0.01 significance level can the manufacturer's claim be refuted? Determine the *p* value. Interpret the result.

Solution 1 – Five-Step Procedure for Testing a Hypothesis

Step 1: *State the null and alternate hypotheses*:

The null hypothesis refers to the "no change" situation. That is, there has been no change in the Clipper's mileage; it is 40 or more mpg.

It is written: H_0: $\mu \geq 40$ and is read that the population mean is greater than or equal to 40.

The alternate hypothesis written: H_1: $\mu < 40$ and is read that the population mean is less than 40.

If the null hypothesis is rejected, then the alternate hypothesis is accepted. It would be concluded that the Clipper's mileage is less than 40 mpg.

Step 2: *Select the level of significance:*

The testing agency decided on the 0.01significance level. This is the probability that the null hypothesis will be rejected when in fact it is true.

Step 3: *Decide on a test statistic:*

The use of the standard normal distribution requires that the population standard deviation σ be known. When it is not known, as in this example, the sample standard deviation designated by s, is used as an estimate of σ. When the sample standard deviation is based on a large sample, the standard normal distribution is still an appropriate test statistic. "Large" is usually defined as being more than 30.

To determine z we use formula [10-2].

$$z = \frac{\overline{X} - \mu}{s / \sqrt{n}}$$

Where:
$\overline{X}$ is the sample mean.
μ is the population mean.
s is the standard deviation computed from the sample.
n is the sample size.

Step 4: *Formulate the decision rule:*

The decision rule is a statement of the conditions under which the null hypothesis is rejected. The decision rule is shown in the following diagram. If the computed value of z is to the left of –2.33, the null hypothesis is rejected. The –2.33 is the critical value. How is it determined?

Remember that the significance level stated in the example is 0.01. This indicates that the area to the left of the critical value under the normal curve is 0.01. For the standard normal distribution the total area to the left of 0 is 0.5000. Therefore, the area between the critical value and 0 is 0.4900, found by 0.5000 – 0.0100. Now refer to Appendix D and search the body of the table for a value close to 0.4900. The closest value is 0.4901. Read 2.3 in the left margin and 0.03 in the column containing 0.4901. Thus the z value corresponding to 0.4901 is 2.33.

Recall from Step 1 that the alternate hypothesis is H_1: $\mu < 40$. The inequality sign points in the negative direction. Thus the critical value is –2.33 and the rejection region is all in the lower left tail.

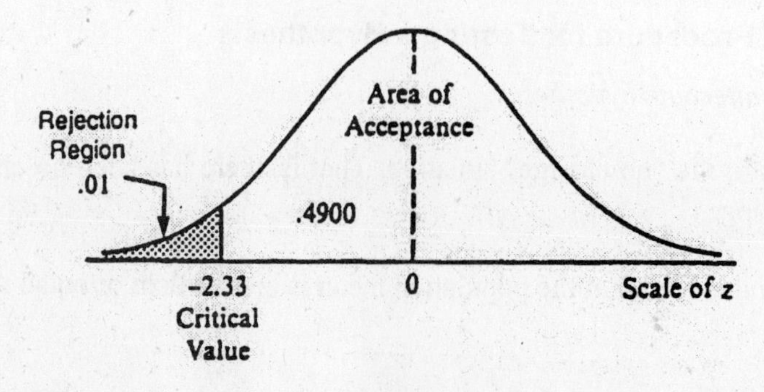

Step 5: *Compute the value of the test statistic, make a decision regarding the null hypothesis, and interpret the results*:

Since the standard deviation of the population is not known, the sample standard deviation is used as its estimate. Repeating the formula for z:

$$z = \frac{\overline{X} - \mu}{s / \sqrt{n}}$$

Recall that the manufacturer claims 40 mpg and the mean of the sample is 38.9 mpg. Solving for z:

$$z = \frac{38.9 - 40.0}{4.0 / \sqrt{64}} = \frac{-1.1}{0.5} = -2.20$$

The computed value of −2.20 is to the right of −2.33, so the null hypothesis is not rejected. We do not reject the claim of the manufacturer that the Clipper gets at least 40.0 miles per gallon. It is reasonable that the 1.1 miles per gallon between (40.0 and 38.9) could be due to chance.

We do observe, however, that −2.20 is fairly close to the critical value of −2.33. What is the likelihood of a z value to the left of −2.20? It is 0.0139, found by 0.5000 – 0.4861, where 0.4861 is the likelihood of a z value between 0 and 2.20. (The probabilities are found in Appendix D.)

The 0.0139 is referred to as the p-value. It is the probability of getting a value of the test statistic (z in this case) more extreme than that actually observed, if the null hypothesis is true. Had the significance level been set at 0.02 instead of .0.01, the null hypothesis would have been rejected. By reporting the p-value we give information on the strength of the decision regarding the null hypothesis.

Self-Review 10.1

Check your answers against those in the ANSWER section.

Last year the records of Dairy Land Inc., a convenience store chain, showed the mean amount spent by a customer was $30. A sample of 40 transactions this month revealed the mean amount spent was $33 with a standard deviation of $12. At the 0.05 significance level, can we conclude that the mean amount spent has increased? What is the p-value? Follow the five-step hypothesis testing procedure.

Example 2 – Five-Step Procedure for Testing a Hypothesis

The mean construction time for a standard two-car garage by Arrowhead Construction Company is 3.5 days. The time for the construction process follows the normal distribution. The construction process is modified through the use of a "quick setting concrete" for the foundation and floor. This should allow the next phase of construction to start in a timelier manner. A sample of 12 garages had a mean construction time of 3.0 days with a standard deviation of 0.9 days. Does use of the quick setting concrete decrease the construction time?

Solution 2 – Five-Step Procedure for Testing a Hypothesis

Step 1: *State the null and alternate hypotheses*:

The null hypothesis is that there is no change in the construction time. That is, the construction time is at least 3.5 days. The alternate hypothesis is that the construction time is less than 3.5 days.

Symbolically, these statements are written as follows:

$$H_0: \mu \geq 3.5$$
$$H_1: \mu < 3.5$$

Step 2: *Select the level of significance:*

We decided on the 0.05 significance level. This is the probability that the null hypothesis will be rejected when in fact it is true.

Step 3: *Select the test statistic:*

The test static in this situation is the t distribution. The distribution of construction times follows the normal distribution, however we do not know the value of the population standard deviation. Also, we have a small sample.

We use text formula [10-5]: $t = \dfrac{\overline{X} - \mu}{s / \sqrt{n}}$

Where:

t is the value of the test statistic.
$\overline{X}$ is the sample mean. (3.0 days)
μ is the population mean. (3.5 days)
s is the standard deviation of the sample. (0.9 days)
n is the sample size. (12)

Thus: $$t = \frac{\overline{X} - \mu}{s / \sqrt{n}} = \frac{3.0 - 3.5}{0.9 / \sqrt{12}} = \frac{-0.5}{0.2598} = -1.925$$

Step 4: *Formulate the decision rule:*

The decision rule is a statement of the conditions under which the null hypothesis is rejected. If the computed value of t is to the left of -1.796, the null hypothesis is rejected. The -1.796 is the critical value. How is it determined?

Remember that the significance level stated in the example is 0.05. The critical values of t are given in Appendix F. The number of degrees of freedom is $(n - 1) = (12 - 1) = 11$. We have a one-tailed test, so we find the portion of the table labeled "one-tailed." Locate the column for the 0.05 significance level. Read down the column until it intersects the row with 11 degrees of freedom. The value is 1.796.

Since this is a one-tailed test and the rejection region is in the left tail, the critical value is negative.

The decision rule is to reject H_0 if the value of t is less than -1.796.

Step 5: *Make a decision regarding the null hypothesis and interpret the results*:

Because −1.925 lies to the left of the critical value −1.796, the null hypothesis is rejected at the 0.05 significance level. This indicates that the use of the "quick setting concrete" has reduced the mean construction time to less than 3.5 days.

> **Self-Review 10.2**
>
> Check your answers against those in the ANSWER section.
>
> The mean construction time for a standard two-car garage by Arrowhead Construction Company is 3.5 days. The time for the construction process follows the normal distribution. The construction process is modified through the use of "precut and assembled roof trusses" rather than onsite construction of roof rafters. This should shorten the construction time. A sample of 15 garages had a mean time of 3.40 days with a standard deviation of 0.8 days. Does use of the "precut and assembled roof trusses" decrease the construction time? Follow the five-step hypothesis testing procedure using the 0.05 significance level.

Example 3 – Using the Hypothesis-Testing Procedure

The Bunting Brass & Bronze Company has a computer controlled machine that is programmed to do precision cutting of a circular brass disc with a mean diameter of 6.125 inches. The shop foreman takes a random sample of 8 discs from the production line. The diameters are as follows:

6.115 6.127 6.129 6.113 6.124 6.121 6.131 6.124

The foreman suspects that the machine is out of adjustment. Use the hypothesis testing procedure to determine if the programmer needs to make adjustments.

Solution 3 – Using the Hypothesis-Testing Procedure

Step 1: *State the null and alternate hypotheses*:

The null hypothesis is that the machine is not out of adjustment. That is, the mean diameter of the discs is 6.125 inches. The alternate hypothesis is that the mean diameter is not 6.125 inches. .

Symbolically, these statements are written as follows:

$$H_0: \mu = 6.125$$
$$H_1: \mu \neq 6.125$$

Step 2: *Select the level of significance:*

We decided on the 0.01 significance level. This is the probability that the null hypothesis will be rejected when in fact it is true.

Step 3: *Select the test statistic:*

The test statistic in this situation is the t distribution. The disc diameters follow the normal distribution, however we do not know the value of the population standard deviation. Also, we have a small sample.

Step 4: *Formulate the decision rule:*

The decision rule is a statement of the conditions under which the null hypothesis is rejected. The alternate hypothesis does not state a direction, so this is a two-tailed test.

Remember that the significance level stated in the example is 0.01. The critical values of t are given in Appendix F. The number of degrees of freedom is $(n - 1) = (8 - 1) = 7$. We have a two-tailed test, so we find the portion of the table labeled "two-tailed." Locate the column for the 0.01 significance level. Read down the column until it intersects the row with 7 degrees of freedom. The value is 3.499.

The decision rule is: Reject the null hypothesis if the computed value of t is to the left of -3.449, or to the right of 3.449

Step 5: *Make a decision regarding the null hypothesis, and interpret the results*:

We use text formula [10-5]:

$$t = \frac{\overline{X} - \mu}{s / \sqrt{n}}$$

Where:

t is the value of the test statistic.
$\overline{X}$ is the sample mean.
μ is the population mean. (6.125 inches)
s is the standard deviation of the sample.
n is the sample size. (8)

We need to calculate the mean and standard deviation of the sample. The standard deviation of the sample can be determined using either formula [3-2] or [3-10].

Computing the sample standard deviation both ways:

X	$X - \overline{X}$	$(X - \overline{X})^2$	X^2
6.115	–0.008	0.000064	37.393225
6.127	0.004	0.000016	37.540129
6.129	0.006	0.000036	37.564641
6.113	–0.010	0.000100	37.368769
6.124	0.001	0.000001	37.503376
6.121	–0.002	0.000004	37.466641
6.131	0.008	0.000064	37.589161
6.124	0.001	0.000001	37.503376
Σ 48.984	0.000	0.000286	299.929318

$$\overline{X} = \frac{\Sigma X}{n} = \frac{48.984}{8} = 6.123$$

$$s = \sqrt{\frac{\Sigma(X - \overline{X})^2}{n-1}} = \sqrt{\frac{0.000286}{8-1}}$$

$$= \sqrt{0.000040857} = 0.0063919 = 0.0064$$

The value of t is computed using formula [10-5]:

$$t = \frac{\overline{X} - \mu}{s / \sqrt{n}} = \frac{6.123 - 6.125}{0.0064 / \sqrt{8}} = \frac{-0.002}{0.0022627} = -0.8839$$

The computed value of –0.8829 lies between the two critical values: –3.449 and 3.449. The null hypothesis is not rejected at the 0.01 significance level. The foreman's suspicion that the machine is out of adjustment cannot be substantiated with this sample.

Self-Review 10.3

Check your answers against those in the ANSWER section.

A typical college student spends an average of 2.55 hours a day using a computer. A sample of 13 students at The University of Findlay revealed the following number of hours per day using the computer:

3.15 3.25 2.00 2.50 2.65 2.75 2.35 2.85 2.95 2.45 1.95 2.35 3.75

Can we conclude that the mean number of hours per day using the computer by students at The University of Findlay is the same as the typical student's usage? Use the hypothesis testing procedure and the 0.05 significance level.

Example 4 – Formulating the Decision Rule

The Dean of Students at Scandia Tech believes that 30 percent of the students are employed. You, as President of the Student Government, believe the proportion employed is less than 30 percent and decide to conduct a study. A random sample of 100 students revealed 25 were employed. At the 0.01 significance level, can the Dean's claim be refuted?

Solution 4 – Formulating the Decision Rule

As usual, the first step is to state the null and alternate hypotheses. The null hypothesis is that there is no change in the percent employed. That is, the population proportion is at least 0.30. The alternate hypothesis is that the population proportion is less than 0.30. This is the statement we are trying to test empirically. Symbolically, these statements are written as follows:

$$H_0: \pi \geq 0.30$$
$$H_1: \pi < 0.30$$

The 0.01 significance level is to be used. The assumptions of the binomial distribution are met in the example. That is

1. There are only two outcomes for each trial--the student is either employed or isn't employed.
2. The number of trials is fixed—100 students.
3. Each trial is independent, meaning the employment of one student selected does not affect another.
4. The probability that any randomly selected student is employed is 0.30.

The normal approximation to the binomial is used because both $n\pi$ and $n(1-\pi)$ exceed 5. That is: $[n\pi = 100(0.30)] = 30$ and $n(1-\pi) = 100(0.70) = 70$. The standard normal distribution, z is the test statistic. To formulate the decision rule, we need the critical value of z. Using the 0.01 significance level, the area is 0.4900, (0.5000 – 0.0100).

Search the body of Appendix D for a value as close to 0.4900 as possible. It is 0.4901. The z value associated with 0.4901 is 2.33. The alternate hypothesis points in the negative direction, hence the rejection region is in the left tail and the critical value of z is –2.33.

The decision rule is to reject the null hypothesis if the computed value of the test statistic lies in the rejection region to the left of –2.33.

Recall that the sample of 100 Scandia Tech students revealed that 25 were employed. The question is whether the sample proportion of 0.25, found by 25/100, is significantly less than 0.30.

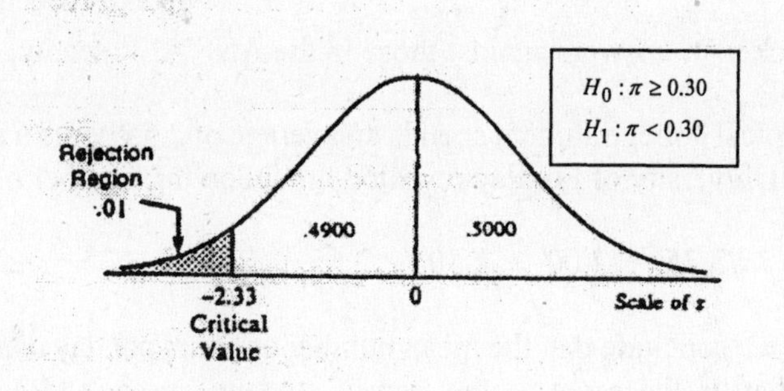

$$z = \frac{p - \pi}{\sqrt{\frac{\pi(1-\pi)}{n}}} = \frac{0.25 - 0.30}{\sqrt{\frac{(0.30)(1-0.30)}{100}}} = -1.09$$

The computed value of z falls in the region between 0 and –2.33. H_0 is not rejected. There is a difference between the Dean's hypothesized proportion (0.30) and the sample proportion (0.25), but this difference of 0.05 is not sufficient to reject the null hypothesis. The 0.05 can be attributed to sampling (chance). The Dean's claim cannot be refuted.

The p-value is the probability of a z value to the left of –1.09. It is 0.1379, found by (0.5000 – 0.3621). The p-value is larger than the significance level of 0.01, which is consistent with our decision not to reject the null hypothesis.

Self-Review 10.4

Check your answers against those in the ANSWER section.

The producer of a TV special expected about 40 percent of the viewing audience to watch a rerun of a 1965 Beatles Concert. A sample of 200 homes revealed 60 to be watching the concert. At the 0.10 significance level, does the evidence suggest that less than 40 percent were watching? Use the usual hypothesis testing format. What is the p-value?

CHAPTER 10 ASSIGNMENT

ONE-SAMPLE TESTS OF HYPOTHESIS

Name ________________________________ Section __________ Score ________

Part I Select the correct answer and write the appropriate letter in the space provided.

______ **1.** The null hypothesis is a claim about
- **a.** the size of the sample.
- **b.** the size of the population.
- **c.** the value of a sample statistic.
- **d.** the value of a population parameter.

______ **2.** When the null hypothesis is rejected, we conclude that
- **a.** the alternate hypothesis is false also.
- **b.** the alternate hypothesis is true.
- **c.** the sample size is too large.
- **d.** we used the wrong test statistic.

______ **3.** A Type I error is committed when
- **a.** *p* value is larger than 1.0.
- **b.** the significance level is greater than 0.05.
- **c.** we reject a true H_0.
- **d.** we accept a false H_0.

______ **4.** The condition or conditions under which H_0 is rejected is
- **a.** called the decision rule.
- **b.** the likelihood of a Type I error.
- **c.** called the test statistic.
- **d.** called the *p*-value.

______ **5.** When the *p*-value is smaller than the significance level
- **a.** a Type I error is committed.
- **b.** a Type II error is committed.
- **c.** the null hypothesis is rejected.
- **d.** the critical value is correct.

______ **6.** A Type II error is
- **a.** rejecting H_1 when it is false.
- **b.** accepting a false H_0.
- **c.** reject H_0 when it is true.
- **d.** not rejecting a false H_1.

______ **7.** In a test regarding a sample mean, σ is not known. Under which of the following conditions can s be substituted for σ and z used as the test statistic?
- **a.** when n is 30 or more.
- **b.** when n is less than 30.
- **c.** when np and $n(1-p)$ both exceed 5.
- **d.** when μ is known.

______ **8.** Under what conditions would a test be considered a one-tailed test.
- **a.** When H_0 contains $\neq$.
- **b.** When there is more than one critical value.
- **c.** When H_1 contains =.
- **d.** When H_1 includes a < or >.

______ **9.** In a two-tailed test the rejection region is
- **a.** all in the upper tail of the standard normal distribution.
- **b.** all in the lower tail of the standard normal distribution.
- **c.** divided equally between the two tails.
- **d.** always equal to −1.96 or 1.96.

______ 10. To compare a single sample proportion to a population proportion

a. n must be less than 30. b. π must be less than 5.

c. $n\pi$ and $n(1-\pi)$ must both be greater than 5. d. σ must be given.

Part II Solve each problem below. Be sure to show essential calculations.

11. The following statements refer to the alternate hypothesis. In the space provided, in symbolic form using H_0, and H_1, write the null and alternate hypothesis.

a. The mean pulse of lawyers is different from 90 beats per minute.

H_0: ______________________________

H_1: ______________________________

b. The mean salary of college presidents is less than $162,500.

H_0: ______________________________

H_1: ______________________________

c. The mean IQ score of 20 year olds is more than 100.

H_0: ______________________________

H_1: ______________________________

d. The mean annual income of sales associates is less than $32,000.

H_0: ______________________________

H_1: ______________________________

12. Identify the Type I and Type II error for each claim in question 11.

a. ______________________________

____

b. ______________________________

____

c. ______________________________

____

d. ______________________________

____

13. A recent article in a computer magazine suggested that the mean time to fully learn a new software program is 40 hours. A sample of 100 first-time users of a new statistics program revealed the mean time to learn it was 39 hours with the standard deviation of 8 hours. At the 0.05 significance level, can we conclude that users learn the package in less than a mean of 40 hours?

   a. State the null and alternate hypotheses.

      H_0: ______________________________

      H_1: ______________________________

   b. State the decision rule.

      ____

   c. Compute the value of the test statistic.

      c.

   d. Compute the p-value.

      d.

   e. What is your decision regarding the null hypothesis? Interpret the result.

      ____

      ____

14. A vinyl siding company claims that the mean time to install siding on a medium-size house is at most 20 hours with a standard deviation of 3.7 hours. A random sample of 40 houses sided in the last three years has a mean installation time of 20.8 hours. At the 0.05 significance level, can a claim be made that it takes longer on average than 20 hours to side a house?

   a. State the null and alternate hypotheses.

      H_0: ______________________________

      H_1: ______________________________

   b. State the decision rule.

      ____

   c. Compute the value of the test statistic.

      c.

d. Formulate the decision rule.

____

e. What is your decision regarding the null hypothesis? Interpret the result.

____

____

15. The mean cleanup and redecorating time for a one-bedroom student apartment at campus Housing is 16 hours. The time for the cleanup and redecorating process follows the normal distribution. The campus Housing administration instituted a "fee and fine system" that encourages students to clean their apartments when they vacate them. This should shorten the cleanup and redecorating time. A sample of 15 apartments had a mean cleanup and redecorating time of 14.5 hours with a standard deviation of 1.5 hours. Does use of the "fee and fine system" decrease the cleanup and redecorating time? Follow the five-step hypothesis testing procedure using the 0.05 significance level.

a. State the null and alternate hypotheses.

H_0: ____

H_1: ____

b. What is the level of significance?

b.

c. Compute the value of the test statistic.

c.

d. Formulate the decision rule.

____

e. What is your decision regarding the null hypothesis? Interpret the result.

____

____

16. A typical college student drinks an average of 96 ounces per day of various beverages that contain caffeine. A sample of 12 students at Wallace College revealed the following amounts of beverages consumed containing caffeine:

108 96 84 84 120 96 108 132 72 120 72 96

Can we conclude that the average amount of beverages consumed containing caffeine at Ownes College is the same as the typical college student? Use the hypothesis testing procedure.

a. State the null and alternate hypotheses.

H_0: ______________________________

H_1: ______________________________

b. State the decision rule.

____

Use the table to

c. & d. Compute the mean and standard deviation.

c.

d.

X	$X-\bar{X}$	$(X-\bar{X})^2$
108		
96		
84		
84		
120		
96		
108		
132		
72		
120		
72		
96		

e. Compute the value of the test statistic.

e.

f. Formulate the decision rule.

____

____

g. What is your decision regarding the null hypothesis? Interpret the result.

____

____

17. CherryBerry Soda, Inc. claims that 15 percent of the population can identify its products. In efforts to boost their identity, CherryBerry employs a famous spokesperson to advertise. A survey is then taken to assess the results of the ad campaign. The results show that 17 percent of the 1000 respondents can identify CherryBerry products. Has the advertising campaign increased product identity or is the difference due to chance? Use the 0.10 significance level.

CHAPTER 11
TWO-SAMPLE TESTS OF HYPOTHESIS

Chapter Goals

After completing this chapter, you will be able to:

1. Conduct a test of a hypothesis about the difference between two independent population means.
2. Conduct a test of a hypothesis about the difference between two population proportions.
3. Conduct a test of a hypothesis about the mean difference between paired or dependent observations.
4. Understand the difference between dependent and independent samples.

Introduction

In this chapter we continue our study of hypothesis testing. Recall that in Chapter 10 we considered hypothesis tests in which we compared the results of a single sample statistic to a population value. When the standard deviation of the population was not known and the sample size was less than 30, we used the Student's *t* distribution.

In this chapter we expand the concept of hypothesis testing to two samples. We select random samples from two different populations and conduct a hypothesis test to determine whether the population means are equal. We might want to test to see if there is a difference in the mean number of defects produced on the 7:00 AM to 3:00 PM shift and the 3:00 PM to 11:00 PM shift at the DaimlerChrysler Jeep Liberty plant in Toledo, Ohio.

Two-Sample Tests of Hypothesis: Independent Samples

As noted above, we expand the concept of hypothesis testing to two samples. When there are two populations, we can compare two sample means to determine if they came from populations with the same or equal means.

For example, a purchasing agent is considering two brands of tires for use on the company's fleet of cars. A sample of 60 Rossford tires indicates the mean useful life to be 65,000 miles. A sample of 50 Maumee tires reveals the useful life to be 68,000 miles. Could the difference between the two sample means be due to chance? The assumption is that for both populations (Rossford and Maumee) the standard deviations are either known or have been computed from samples greater than 30. The test statistic follows the standard normal distribution and its value is computed from text formula [11-2]:

Test Statistic for the Difference Between Two Means

$$z = \frac{\bar{X}_1 - \bar{X}_2}{\sqrt{\frac{s_1^2}{n_1} + \frac{s_2^2}{n_2}}} \quad [11-2]$$

Where:

$\overline{X}_1$ and $\overline{X}_2$ refer to the two sample means.

s_1^2 and s_2^2 refer to the two sample variances.

n_1 and n_2, refer to the two sample sizes.

The following are assumptions necessary for this two-sample test of means:

1. The two populations must be unrelated; that is, independent.
2. The samples must be large enough such that the distribution of the sample means follows the normal distribution. The usual practice is to require that both samples have at least 30 observations.

Two-Sample Tests about Proportions

We are often interested in whether two sample proportions came from populations that are equal. For example, we want to compare the proportion of rural voters planning to vote for the incumbent governor with the proportion of urban voters. The test statistic is formula [11-3]:

Two-Sample Test of Proportions

$$z = \frac{p_1 - p_2}{\sqrt{\frac{p_c(1-p_c)}{n_1} + \frac{p_c(1-p_c)}{n_2}}} \qquad [11-3]$$

Where:

p_1 is the proportion in the first sample possessing the trait.

p_2 is the proportion in the second sample possessing the trait.

n_1 is the number of observations in the first sample.

n_2 is the number of observations in the second sample.

p_c is the pooled proportion possessing the trait in the combined samples. It is called the pooled estimate of the population proportion and is found by formula [11-4]

Pooled Proportion

$$p_c = \frac{X_1 + X_2}{n_1 + n_2} \qquad [11-4]$$

Where:

X_1 is the number possessing the trait in the first sample.

X_2 is the number possessing the trait in the second sample.

Comparing Population Means with Small Samples

We now consider the case in which the population standard deviations are unknown and the number of observations in at least one of the samples is less than 30. This is often referred to as a "small sample test of means."

Three assumptions are required:

1. The sampled populations follow the normal distribution.
2. The two samples are from independent populations.
3. The standard deviations of the two populations are equal.

The t distribution is the test statistic for a test of hypothesis for the difference between two population means. The t statistic for the two sample cases is similar to that employed for the z statistic, with one additional calculation. The two sample variances must be "pooled" to form a single estimate of the unknown population variance.

This is accomplished by using text formula [11-5]:

Pooled Variance

$$s_p^2 = \frac{(n_1 - 1)s_1^2 + (n_2 - 1)s_2^2}{n_1 + n_2 - 2} \quad [11-5]$$

Where:

s_p^2 is the pooled estimate of the population variance.

s_1^2 is the variance (standard deviation squared) of the first sample.

s_2^2 is the variance of the second sample.

n_1 is the number of observations in the first sample.

n_2 is the number of observations in the second sample.

The value of t is then computed using text formula [11-6].

Two-Sample Test of Means—Small Samples

$$t = \frac{\bar{X}_1 - \bar{X}_2}{\sqrt{s_p^2\left(\frac{1}{n_1} + \frac{1}{n_2}\right)}} \quad [11-6]$$

Where:

$\bar{X}_1$ is the mean of the first sample.

$\bar{X}_2$ is the mean of the second sample.

n_1 is the number of observations in the first sample.

n_2 is the number of observations in the second sample.

s_p^2 is the polled estimate of the population variance.

The number of degrees of freedom for a two-sample test is the total number of items sampled minus the number of samples. It is found by: $(n_1 + n_2 - 2)$.

Two-Sample Tests of Hypothesis: Dependent Samples

Another hypothesis testing situation occurs when we are concerned with the difference in paired or related observations. These are situations in which the samples are not independent. Typically, it is a before-and-after situation, where we want to measure the difference.

To illustrate, suppose we administer a reading test to a sample of ten students. We have them take a course in speed reading and then we test them again. Thus the test focuses on the reading improvements of each of the ten students. The distribution of the population of differences is assumed to be approximately normal. The test statistic is t, and text formula [11-7] is used.

Paired *t* Test	$t = \frac{\bar{d}}{s_d / \sqrt{n}}$	[11–7]

Where:

$\bar{d}$ is the mean of the difference between the paired or related observations.
s_d is the standard deviation of the differences between the paired or related observations.
n is the number of paired observations.

For a paired difference test there are $(n - 1)$ degrees of freedom.

The standard deviation of the differences s_d is computed using the familiar formula for the standard deviation except that d is substituted for X. The text formula is: $s_d = \sqrt{\frac{\Sigma(d - \bar{d})^2}{n-1}}$

Comparing Dependent and Independent Samples

When working with paired data, we need to distinguish between ***dependent samples*** and ***independent samples***.

Dependent samples: Two samples that are related to each other.

There are two types of dependent samples:

1. *Samples characterized by a measurement, an intervention of some type, and then another measurement.* This is often referred to as a "before" and "after" study. For example: A group of teenagers are enrolled in a weight reduction program. They are weighted, go through a diet and exercise program, and then they are weighed again. The two weights are paired weights and are considered to be dependent samples. The paired samples are dependent because the same individual is a member of both samples.

2. *Samples characterized by matching or pairing observations.* For example: The transportation manager wants to study the amount of "tire wear" on two brands of tire. One tire of each brand is placed on 15 company trucks and the wear is measured after 20,000 miles. The manager would have 30 observations with 15 pairs of data. The paired samples are dependent because the pairs came off the same truck.

When the samples chosen at random are in no way related to each other they are considered independent samples.

Independent samples: Two samples which are unrelated to each other.

Independent samples are essentially samples taken from entirely different populations. Keep in mind, however that the populations need to share some similar characteristics. For example: The human resource director might want to test the "Microsoft Word" skills of two sets of graduates from two different secretarial business programs.

Glossary

Dependent samples: Two samples that are related to each other.

Independent samples: Two samples which are unrelated to each other.

Chapter Examples

Example 1 – Two-Sample Test of Hypothesis: Independent Samples

Two manufacturers of sinus relief tablets, SINUS and ANTIDRIP, have made conflicting claims regarding the effectiveness of their tablets. A private testing organization was hired to evaluate the two tablets. The testing company tried SINUS on 100 sinus congestion sufferers and found the mean time to relief was 85.0 minutes with a sample standard deviation of 6.0 minutes. A sample of 81 sinus congestion sufferers used ANTIDRIP. The mean time to relief was 86.2 minutes, the sample standard deviation 6.8 minutes. Does the evidence suggest a difference in the amount of time required to obtain relief? Use the 0.05 significance level and the five-step procedure. What is the p-value? Interpret it.

Solution 1 – Two-Sample Test of Hypothesis: Independent Samples

We use the five-step hypothesis testing procedure for the solution.

Step 1: State the null hypothesis and the alternate hypothesis.

Note that the testing company is attempting to show only that there is a difference in the time required to affect relief. There is no attempt to show one tablet is "better than" or "worse than" the other. Thus, a two-tailed test is applied.

$$H_0: \mu_1 = \mu_2$$
$$H_1: \mu_1 \neq \mu_2$$

Where: μ_1 refers to the mean time to obtain relief using SINUS

μ_2 refers to the mean time to obtain relief using ANTIDRIP.

Step 2: Select a level of significance.

The 0.05 significance level is to be used. The alternate hypothesis does not state a direction, so this is a two-tailed test. The 0.05 significance level is divided equally into two tails of the standard normal distribution. Hence, the area in the left tail is 0.0250 and the area in the right tail is 0.0250.

Step 3: Identify the test statistic.

Because both samples are large (greater than 30) the z distribution is used as the test statistic.

Step 4: Formulate a decision rule based on the selected test statistic and level of significance.

The critical values that separate the two rejection regions from the region of acceptance are –1.96 and +1.96. To explain: if the area in a rejection region is 0.0250, the acceptance area is 0.4750, found by 0.5000 – 0.0250. The z value corresponding to an area of 0.4750 is obtained by referring to the table of areas of the normal curve (Appendix D). Search the body of the table for a value as close to 0.4750 as possible and read

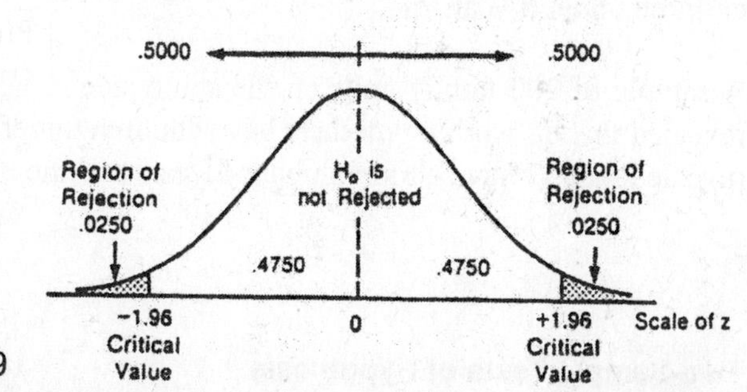

the corresponding row and column values. The area of 0.4750 is found in the row 1.9 and the column 0.06. Hence, the critical values are + 1.96 or −1.96.

This decision rule is shown on the diagram at the right.

Step 5: Make a decision to reject or not to reject the null hypothesis and interpret the results.

The computed value of z is −1.24, found by using formula [11-2]. Because the population standard deviations are not known, the sample standard deviations are substituted.

$$z = \frac{\overline{X}_1 - \overline{X}_2}{\sqrt{\frac{s_1^2}{n_1} + \frac{s_2^2}{n_2}}} = \frac{85.0 - 86.2}{\sqrt{\frac{(6.0)^2}{100} + \frac{(6.8)^2}{81}}} = -1.24$$

The computed value of z is between −1.96 and +1.96. Thus H_0 is not rejected. We conclude that there is no difference in the mean time it takes SINUS and ANTIDRIP to bring relief. The difference of 1.2 minutes (85.0 – 86.2) can be attributed to sampling error (chance).

To determine the p-value we need to find the area to the left of −1.24 and add to it the area to the right of 1.24. We are concerned with both tails because H_1 is two-tailed. The p-value is 0.2150, found by 2(0.5000 – 0.3925). Since the p-value of 0.2150 is greater than the level of significance of 0.05, do not reject H_0.

Self-Review 11.1

Check your answers against those in the ANSWER section.

The county commissioners received a number of complaints from county residents that the Youngsville Fire Department takes longer to respond to emergency runs than the Claredon Fire Department. To check the validity of these complaints, a random sample of 60 emergency runs handled by the Youngsville Fire Department was selected. It was found that the mean response time was 6.9 minutes and the standard deviation of the sample 3.8 minutes. A sample of 70 emergency runs handled by the Claredon Fire Department found the mean response time was 4.9 minutes with a sample standard deviation of 3.0 minutes. Does the data suggest that it takes longer for the Youngsville Department to respond? Use the 0.05 significance level.

Example 2 – Two-Sample Tests About Proportion

Two different sites are being considered for a day-care center. One is on the south side of the city and the other is on the east side. The decision where to locate the day-care center depends in part on how many mothers work and have children under 5 years old.

	South Side	East Side
Number of working mothers with children under 5	$X_1 = 88$	$X_2 = 57$
Number in sample	$n_1 = 200$	$n_2 = 150$
Proportion with children under 5 and mothers work	$p_1 = 0.44$	$p_2 = 0.38$

A sample of 200 family units on the south side revealed that 88 working mothers have children under 5 years. A sample of 150 family units on the east side revealed that 57 have children under 5 years and the mother worked. Summarizing the data above:

Can we conclude that in the population a larger proportion of mothers on the south side work and have children under 5 than on the east side? Or, can the difference be attributed to sampling variation (chance)? Use the 0.05 level of significance.

Solution 2 – Two-Sample Tests About Proportion

The problem is to examine whether a higher proportion of working mothers of young children live on the south side.

Step 1: State the null hypothesis and the alternate hypothesis.

The hypotheses are:

$$H_0 : \pi_1 \leq \pi_2$$
$$H_1 : \pi_1 > \pi_2$$

Where:

π_1 refers to the proportion of working mothers on the south side.

π_2 refers to the proportion of working mothers on the east side.

Step 2: Select a level of significance.

The 0.05 significance level is stated in the **Example**.

Step 3: Identify the test statistic.

The standard normal distribution is the test statistic to be used. The z value is computed using formula [11-3].

Step 4: Formulate a decision rule based on the selected test statistic and level of significance.

The alternate hypothesis indicates a direction, so this is a one-tailed test.

The critical value is 1.65 obtained from Appendix D. The area in the upper tail of the curve is 0.05, therefore the area between $z = 0$ and the critical value is 0.4500, found by (0.5000 – 0.0500). Search the body of the table for a value close to 0.4500. Since 1.64 is equal to 0.4495 and 1.65 is equal to 0.4505, a value between 1.64 and 1.65 or (1.645) could be used as the critical value. We take the conservative approach and use 1.65. The null hypothesis is rejected if the calculated z value is greater than 1.65. This information is summarized in the diagram above.

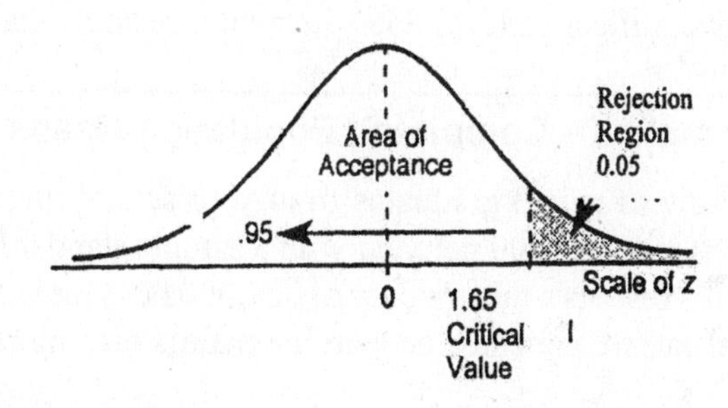

Formula [11-3] for z is repeated below.

$$z = \frac{p_1 - p_2}{\sqrt{\frac{p_c(1-p_c)}{n_1} + \frac{p_c(1-p_c)}{n_2}}}$$

where p_c is a pooled estimate of the population proportion and is computed using formula [11-4].

$$p_c = \frac{X_1 + X_2}{n_1 + n_2}$$

In this **Example** X_1, and X_2 refer to the number of "successes" in each sample (number of working mothers with children under 5 years), n_1 and n_2 refer to the number of housing units sampled in the south side and east side, respectively. The pooled estimate of the population proportion is 0.4143, found as follows:

$$p_c = \frac{X_1 + X_2}{n_1 + n_2} = \frac{88 + 57}{200 + 150} = 0.4143$$

Inserting the pooled estimate of 0.4143 in the formula and solving for z in formula [11-3] gives 1.13.

$$z = \frac{p_1 - p_2}{\sqrt{\frac{p_c(1-p_c)}{n_1} + \frac{p_c(1-p_c)}{n_2}}} = \frac{0.44 - 0.38}{\sqrt{\frac{(0.4143)(1-0.4143)}{200} + \frac{(0.4143)(1-0.4143)}{150}}} = 1.13$$

Step 5: Make a decision to reject or not to reject the null hypothesis.

The computed value: $z = 1.13$, is less than the critical value of 1.65 so the null hypothesis is not rejected. The difference can be attributed to sampling error (chance). To put it another way, the proportion of mothers who work and have children under 5 on the south side is not significantly greater than the east side. The *p*-value is 0.1292, found by (0.5000 – 0.3708). So, the probability of finding a value of the test statistic this large or larger is 0.1292.

Self-Review 11.2

Check your answers against those in the ANSWER section.
A recent study was designed to compare smoking habits of young women with those of young men. A random sample of 150 women revealed that 45 smoked. A random sample of 100 men indicated that 25 smoked. At the 0.05 significance level does the evidence show that a higher proportion of women smoke? Compute the *p*-value.

Example 3 – Comparing Population Means with Small Samples

A study of recent graduates from your school reveals that for a sample of ten accounting majors the mean salary was $30,000 per year with a sample standard deviation of $2,000. A sample of eight general business majors reveals a mean salary of $29,000 per year with a standard deviation of $1,500. At the 0.05 significance level can we conclude accounting majors earn more?

Solution 3 – Comparing Population Means with Small Samples

The null hypothesis is that accounting majors earn the same or less than general business majors. The alternate hypothesis is that accounting majors earn more. They are written as follows:

$$H_0: \mu_1 \leq \mu_2$$
$$H_1: \mu_1 > \mu_2$$

Where:
μ_1 refers to accounting majors (graduates).
μ_2 refers to general business majors (graduates).

The required assumptions are:

1. The samples are independent.
2. The two populations follow the normal distribution.
3. The population standard deviations are equal.

The t distribution is the test statistic. There are 16 degrees of freedom, found by $(n_1 + n_2 - 2) = (10 + 8 - 2) = 16$. The alternate hypothesis is a one-tailed test with the rejection region in the upper tail. From Appendix F, the critical value is 1.746. Hence, H_0 is rejected if the computed value of the test statistic exceeds 1.746.

The first step is to pool the variances, using formula [11-5].

$$s_p^2 = \frac{(n_1 - 1)(s_1^2) + (n_2 - 1)(s_2^2)}{n_1 + n_2 - 2} = \frac{(10-1)(2{,}000)^2 + (8-1)(1{,}500)^2}{10+8-2} = 3{,}234{,}375$$

Next, the value of t is computed, using formula [11-6].

$$t = \frac{\overline{X}_1 - \overline{X}_2}{\sqrt{s_p^2\left(\frac{1}{n_1} + \frac{1}{n_2}\right)}} = \frac{\$30{,}000 - \$29{,}000}{\sqrt{(3{,}234{,}375)\left(\frac{1}{10} + \frac{1}{8}\right)}} = \frac{1{,}000}{\sqrt{3{,}234{,}375(0.225)}} = \frac{1{,}000}{853.073} = 1.17$$

Because the computed value of t (1.17) is less than the critical value of 1.746, H_0 is not rejected. The sample evidence does not suggest a difference in the mean salaries of the two groups.

We determine the p-value by using Appendix F. Move down the left column to the 16 degrees of freedom row. Move across the row until you locate a value larger than 1.17—the computed value of t. The value is in the first column. It is 1.337. Note that this column has 0.10 as a heading. Since this is a one-tailed test, we conclude that the p-value is greater than 0.10.

Self-Review 11.3

Check your answers against those in the ANSWER section.

A large department store hired a researcher to compare the average purchase amounts for the downtown store with that of its mall store. The information shown was obtained. At the 0.01 significance level can it be concluded that the mean amount spent at the mall store is larger? Estimate the p-value.

	Downtown Store	**Mall Store**
Average purchase amount	$36.00	$40.00
Sample standard deviation	$10.00	$12.00
Sample size	10	10

Example 4 – Two-Sample Test of Hypothesis: Dependent Samples

The Dean of the College of Business at Kingsport University wants to determine if the Grade Point Average (GPA) of business college students decreases during the last semester of their senior year. A sample of six students is selected. Their GPAs for the fall and spring semesters of their senior year are:

Student	**Fall Semester**	**Spring Semester**
A	2.7	3.1
B	3.4	3.3
C	3.5	3.3
D	3.0	2.9
E	2.1	1.8
F	2.7	2.4

At the 0.05 significance level, can the Dean conclude that the GPA of graduating seniors declined during their last semester?

Solution 4 – Two-Sample Test of Hypothesis: Dependent Samples

Let μ_d be the mean difference between the fall and spring semester grades for all business students at Kingsport U. in their senior year. A one-tailed test is appropriate since we want to explore whether grades decrease.

$$H_0: \mu_d \leq 0$$
$$H_1: \mu_d > 0$$

There are six paired observations; therefore, there are $(n - 1) = (6 - 1) = 5$ degrees of freedom. Using Appendix F with 5 degrees of freedom, the 0.05 significance level and a one-tailed test, the critical value of t is 2.015. H_0 is rejected if the computed value of the test statistic exceeds 2.015.

The value of the test statistic is determined from formula [11-7]. $t = \dfrac{\bar{d}}{s_d / \sqrt{n}}$

Where:

$\bar{d}$ is the mean of the differences between fall and spring GPAs.
s_d is the standard deviation of those differences.
n is the number of paired observations.

First, subtract Spring semester grades from fall semester grades. If this difference is positive, then a decline has occurred. The sample data is shown below and the values of d and s_d computed:

Student	Fall	Spring	d	$(d - \bar{d})$	$(d - \bar{d})^2$
A	2.7	3.1	−0.4	−0.5	0.25
B	3.4	3.3	0.1	0	0
C	3.5	3.3	0.2	0.1	0.01
D	3.0	2.9	0.1	0	0
E	2.1	1.8	0.3	0.2	0.04
F	2.7	2.4	0.3	0.2	0.04
			0.6		0.34

$$\bar{d} = \frac{\Sigma d}{N} = \frac{0.6}{6} = 0.10$$

$$s_d = \sqrt{\frac{\Sigma(d - \bar{d})^2}{n-1}} = \sqrt{\frac{0.34}{6-1}} = 0.2608$$

The t statistic is computed by: $t = \dfrac{\bar{d}}{s_d / \sqrt{n}} = \dfrac{0.10}{0.2608 / \sqrt{6}} = \dfrac{0.10}{0.1065} = 0.94$

Since the computed value of t (0.94) is less than the critical value of 2.015, H_0 is not rejected. The evidence does not suggest a reduction in grades from the fall to the spring semester. The decrease in GPAs can be attributed to chance. The p-value is greater than 0.10.

Self-Review 11.4

Check your answers against those in the ANSWER section.

An independent government agency is interested in comparing the heating cost of all- electric homes and those of homes heated with natural gas. A sample of eight all-electric homes is matched with eight homes of similar size and other features that use natural gas. The heating costs for last January are obtained for each home.

At the 0.05 significance level is there reason to believe there is a difference in heating costs?

Matched Pair	Electric Heat	Gas Heat
1	265	260
2	271	270
3	260	250
4	250	255
5	248	250
6	280	275
7	257	260
8	262	260

CHAPTER 11 ASSIGNMENT

TWO-SAMPLE TESTS OF HYPOTHESIS

Name ______________________ Section __________ Score ________

Part I Select the correct answer and write the appropriate letter in the space provided.

_____ **1.** The test statistic for testing a hypothesis for large sample means when the population standard deviation is not known is

a. the t distribution. **b.** the F distribution.
c. the z distribution. **d.** the μ distribution.

_____ **2.** We want to test a hypothesis for the difference between two population means. The null and alternate hypothesis are indicated:

$$H_0: \pi_1 \leq \pi_2$$
$$H_1: \pi_1 > \pi_2$$

a. A left-tailed test should be applied
b. A right-tailed test should be applied
c. A two-tailed test should be applied
d. We cannot determine whether a left, right or two-tailed test to apply without more information

_____ **3.** In a large sample test of means, both samples must have at least:

a. 120 items **b.** 90 items
c. 60 items **d.** 30 items.

_____ **4.** In a two-sample test of means for independent samples, $n_1 = 12$ and $n_2 = 10$. How many degrees of freedom are in the test?

a. 22 **b.** 21
c. 20 **d.** none of the above

_____ **5.** In the paired t-test, we assume in the null hypothesis that the distribution of the differences between the paired observation has a mean

a. equal to 1. **b.** equal to n - 1.
c. equal to 0. **d.** none of the above

_____ **6.** For a particular significance level and sample size the value of the t for a one-tailed test is

a. always less than z. **b.** always more than z.
c. equal to 0. **d.** equal to z.

_____ **7.** Which of the following is **not** an assumption for the two-sample t-test?

a. equal *sample* variances **b.** independent samples
c. normal populations **d.** equal *population* standard deviations

_____ 8. For dependent samples, we assume the distribution of the differences in the populations has a mean of:

a. 30 b. 0
c. 25 d. none of the above

_____ 9. For tests of hypothesis for a two-sample means, sample size greater than 30, a one-tailed test (rejection region in the upper tail), using the 0.01 significance level, the critical value is:

a. 2.18 b. 2.68
c. 2.33 d. 3.12

_____ 10. To determine if a diet supplement is useful for increasing weight, patients are weighed at the start of the program and at the end of the program. This is an example of a(n)

a. test of paired differences. b. independent sample.
c. one-sample test for means. d. two-sample test for means.

Part II Answer the following questions. Be sure to show essential work.

11. A financial planner wants to compare the yield of income- and growth-oriented mutual funds. Fifty thousand dollars is invested in each of a sample of 35 income-oriented and 40 growth-oriented funds. The mean increase for a two-year period for the income funds is $1100 with a standard deviation of $45. For the growth-oriented funds the mean increase is $1090 with a standard deviation of $55. At the 0.01 significance level is there a difference in the mean yield of the two funds?

a. State the null and alternate hypotheses.

H_0: ____________________ H_1: ____________________

b. State the decision rule.

____

c. Compute the value of the test statistic.

c.

d. Compute the p-value.

d.

e. What is your decision regarding the null hypothesis?

____

12. Is the mean salary of accountants who have reached partnership status higher than that for accountants who are not partners? A sample of 15 accountants who have the partnership status showed a mean salary of $82,000 with a standard deviation of $5,500. A sample of 12 accountants who were not partners showed a mean of $78,000 with a standard deviation of $6,500. At the 0.05 significance level can we conclude that accountants at the partnership level earn larger salaries?

a. State the null and alternate hypotheses.

H_0: ______________________ H_1: ______________________

b. State the decision rule.

____

c. Compute the value of the test statistic.

c.

d. Compute the p-value.

d.

e. What is your decision regarding the null hypothesis?

____

13. The Human Resources Director for a large company is studying absenteeism among hourly workers. A sample of 120 day shift employees showed 15 were absent more than five days last year. A sample of 80 afternoon employees showed 18 to be absent five or more times. At the 0.01 significance level can we conclude that there is a higher proportion of absenteeism among afternoon employees?

a. State the null and alternate hypotheses.

H_0: ______________________ H_1: ______________________

b. State the decision rule.

____

c. Compute the value of the test statistic.

c.

d. Compute the p-value.

d.

e. What is your decision regarding the null hypothesis?

____

14. The President and CEO of Cliff Hanger International Airlines is concerned about high cholesterol levels of the pilots. In an attempt to improve the situation a sample of seven pilots is selected to take part in a special program, in which each pilot is given a special diet by the company physician. After six months each pilot's cholesterol level is checked again. At the 0.01 significance level can we conclude that the program was effective in reducing cholesterol levels?

Pilot	Before	After	d	$(d - \bar{d})$	$(d - \bar{d})^2$
1	255	210			
2	230	225			
3	290	215			
4	242	215			
5	300	240			
6	250	235			
7	215	190			

a. State the null and alternate hypotheses.

H_0: ____________________ H_1: ____________________

b. State the decision rule.

____

c. Compute the value of the test statistic.

c.

d. Compute the p-value.

d.

e. What is your decision regarding the null hypothesis?

____

CHAPTER 12
ANALYSIS OF VARIANCE

Chapter Goals

After completing this chapter, you will be able to:

1. List the characteristics of the *F* distribution.
2. Conduct a test of hypothesis to determine whether the variances of two populations are equal.
3. Discuss the general idea of analysis of variance.
4. Organize data into a one-way ANOVA table.
5. Conduct a test of hypothesis among three or more treatment means.
6. Develop confidence intervals for the difference in treatment means.

Introduction

In Chapter 10 we developed a method to determine whether there is a difference between two population means when the sample sizes are each less than 30. What if we wanted to compare more than two populations? The two-sample *t* test used in Chapter 10 requires that the populations be compared two at a time. This would be very time consuming, would offer the possibility of errors in calculations, but most seriously there would be a build-up of Type I error. That is, the total value of α would become quite large as the number of comparisons increased. In this chapter we will describe a technique that is efficient when simultaneously comparing several sample means to determine if they come from the same or equal populations. This technique is known as ***Analysis of Variance (ANOVA).***

Analysis of Variance (ANOVA). A statistical technique used to determine whether several populations have the same mean. This is accomplished by comparing the sample variances.

A second test compares two sample variances to determine if the populations are equal. This test is particularly useful in validating a requirement of the two-sample *t* tests that both populations have the same standard deviations.

The *F* Distribution

The test statistic used to compare the sample variances and to conduct the ANOVA test is the ***F distribution***.

F Distribution. A continuous probability distribution where *F* is always 0 or positive. The distribution is positively skewed. It is based on two parameters, the number of degrees of freedom in the numerator and the number of degrees of freedom in the denominator.

The major characteristics of the *F* distribution are:

1. **There is a "family" of *F* distributions**. A particular member of the family is determined by two parameters: the degrees of freedom in the numerator and the degrees of freedom in the denominator.

2. **The F distribution is continuous.** This means that it can assume an infinite number of values between 0 and positive infinity.

3. **The F distribution cannot be negative.** The smallest value F can assume is 0.

4. **It is positively skewed.** The long tail of the distribution is to the right-hand side. As the number of degrees of freedom increases in both the numerator and denominator, the distribution approaches a normal distribution.

5. **It is asymptotic.** As the values of X increase, the F curve approaches the X-axis but never touches it. This is similar to the behavior of the normal distribution described in Chapter 7.

Comparing Two Populations Variances

The F distribution is used to test the hypothesis that the variance of one normal population equals the variance of another normal population. The F distribution can also be used to validate assumptions with respect to certain statistical tests. Regardless of whether we want to determine if one population has more variation than another population does or validate an assumption with respect to a statistical test, we still use the usual five-step hypothesis testing procedure. The value of the test statistic is determined using text formula [12-1].

Test Statistic for Comparing Two Variances

$$F = \frac{s_1^2}{s_2^2} \qquad [12-1]$$

Where:

s_1^2 is the variance of the first sample.

s_2^2 is the variance of the second sample.

The usual practice is to determine the F ratio by putting the larger of the two sample variances in the numerator. This will force the F ratio to be larger than 1.00. This allows us to always use the upper tail of the F statistic, thus avoiding the need for more extensive F tables.

Again, the F distribution is used to determine if the sample variance from one normal population is the same as the variance obtained from another normal population.

For example, if you were comparing the mean starting salaries for this year's marketing graduates to this year's computer science graduates, an assumption required of the two-sample t test is that both populations have the same standard deviation. Therefore, before conducting the test for means, it is essential to show that the two population variances are equal.

The idea behind the test for standard deviations is that if the null hypothesis is true that the two sample variances are equal, then their ratio will be approximately 1.00. If the null hypothesis is false, then the ratio will be much larger than 1.00. The F distribution provides a decision rule to let us know when the departure from 1.00 is too large to have happened by chance.

ANOVA Assumptions

Another use of the F distribution is the analysis of variance (ANOVA) technique where we compare three or more population means to determine whether they could be equal. To employ ANOVA, three conditions must be met:

1. The populations follow the normal distribution.
2. The populations have equal standard deviations (σ).
3. The populations are independent.

When these conditions are met, F is used as the test statistic to measure the variance among means.

The ANOVA Test

The ANOVA test is used to determine if the various sample means came from a single population or populations with different means. The sample means are compared through their variances. The underlying strategy is to estimate the population variance two ways and then find the ratio of these two estimates. If the ratio is about one (1), then the two estimates are the same and we conclude that the population means are the same. If the ratio is quite different from 1, then we conclude that the population means are not the same. The F distribution tells when the ratio is too much larger than 1 to have occurred by chance.

The same hypothesis testing procedure used with the standard normal distribution (z) and Student's t is also employed with analysis of variance. The test statistic is the F distribution.

Step 1. State the null hypothesis and the alternate hypothesis.

When three population means are compared, the null and alternate hypotheses are written:

$$H_0 : \mu_1 = \mu_2 = \mu_3$$
$$H_1 : \text{not all means are equal}$$

Note that rejection of the null hypothesis does not identify which populations differ significantly. It merely indicates that a difference between at least one pair of means exists.

Step 2. Select the level of significance.

The most common values selected are 0.01 or 0.05.

Step 3. Identify the test statistic.

For an analysis of variance problem the appropriate test statistic is F. The F statistic is the ratio of two variance estimates and is computed by the formula:

$$F = \frac{\text{Estimate of the population variance based on the differences among the sample means}}{\text{Estimate of the population variance based on the variation within samples}}$$

There are $(k-1)$ degrees of freedom associated with the numerator of the formula for F, and $(n-k)$ degrees of freedom associated with the denominator, where k is the number of populations and n is the total number of sample observations.

Step 4. Formulate the Decision Rule.

The critical value is determined from the F table found in Appendix G.

To illustrate how the decision rule is established, suppose a package delivery company purchased 14 trucks at the same time. Five trucks were purchased from Ford, four from General Motors (GM), and five from DaimlerChrysler. All the trucks were used to deliver packages. The cost of maintaining the trucks for the first year is shown. Is there a significant difference in the mean maintenance cost of the three manufacturers?

Maintenance Cost, By Manufacturer		
Ford	**Daimler Chrysler**	**GM**
$ 914	$933	$1,004
1,000	874	1,114
1,127	927	1,044
988	983	1,100
947		1,139

The three different manufacturers are called ***treatments.***

> ***Treatments.*** A specific source of variation in a set of data.

The term is borrowed from agricultural research, where much of the early development of the ANOVA technique took place. Crop yields were compared after different fertilizers (that is, treatments) had been applied to various plots of land.

In the study comparing truck manufacturers there are three treatments. Therefore there are two degrees of freedom in the numerator, found by $(k - 1) = (3 - 1) = 2$. How is the number of degrees of freedom for the denominator determined? Note that in the three samples there are a total of 14 observations. Thus the total number of observations, designated by n, is 14. The number of degrees of freedom in the denominator is 11, found by $(n - k) = (14 - 3) = 11$.

The critical value of F can be found in Appendix G at the back of the study guide. There are tables for both the 0.01 and the 0.05 significance levels. Using the 0.05 significance level, the degrees of freedom for the numerator are at the top of the table and for the denominator in the left margin. To locate the critical value, move horizontally at the top of the table to 2 degrees of freedom in the numerator, then down that column to the number opposite 11 degrees of freedom in the left margin (denominator). That number is 3.98, which is the critical value of F.

The decision rule is to reject the null hypothesis if the computed value of F exceeds 3.98; otherwise it is not rejected. To reject the null hypothesis and accept the alternate hypothesis allows us to conclude that there is a significant difference between at least one pair of means. If the null hypothesis is not rejected, this implies the differences between the sample means could have occurred by chance. Portrayed graphically, the decision rule is shown at the right.

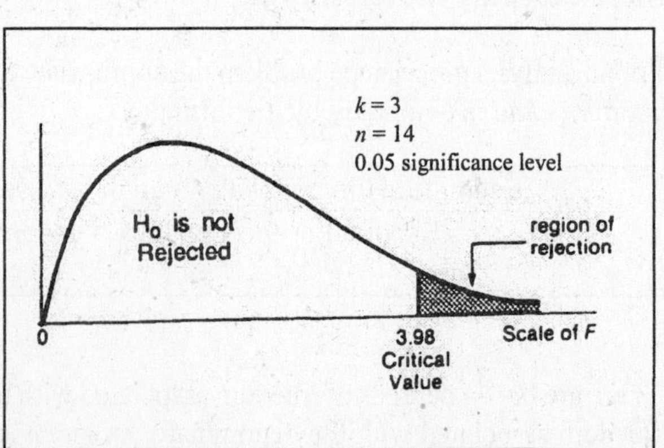

Step 5. Select the sample, perform the calculations, and make a decision.

The value of F is computed from the sample information and a decision is made regarding the null hypothesis. If the computed value of F is 1.98, for example, the null hypothesis is not rejected. If it is greater than 3.98, say 4.26, then the null hypothesis is rejected and the alternate accepted.

The ANOVA Table

A convenient way of organizing the calculations for F is to put them in a table referred to as an ANOVA table.

ANOVA Table				
Source of Variation	**Sum of Squares**	**Degrees of Freedom**	**Mean Square**	***F***
Treatments	SST	$k-1$	SST/ $(k-1)$ = MST	MST / MSE
Error	SSE	$n-k$	SSE/ $(n-k)$ = MSE	
Total	SS Total	$n-1$		

Note in the ANOVA Table that there are three values, called **Sum of Squares**, that are required to compute F. The three values are determined by first calculating SS total and SSE, then finding SST by subtraction. Keep in mind the fact that the SS total term is the total variation, SST is the variation due to the treatments, and SSE is the variation within the treatments.

We work across the table to eventually calculate F. The degrees of freedom are the same as those used to find the critical value of F. The term **mean square** is another expression for estimate of the variance. As seen in the table, the **mean square for treatments** (written MST) is SST divided by the degrees of freedom. Also from the table the **mean square error** (written MSE) is SSE divided by the degrees of freedom. To find F we divide MST by MSE. The entire process is explained in Example 1.

Glossary

Analysis of Variance (ANOVA): A statistical technique used to determine whether several populations have the same mean. This is accomplished by comparing the sample variances.

***F Distribution*:** A continuous probability distribution where F is always 0 or positive. The distribution is positively skewed. It is based on two parameters, the number of degrees of freedom in the numerator and the number of degrees of freedom in the denominator.

***Treatments*:** A treatment is a specific source, or cause, of variation in a set of data.

Chapter Examples

Example 1 – Comparing Two Populations Variances Using a Two-Tailed Test

Teledko Associates is a marketing research firm that specializes in comparative shopping. Teledko is hired by General Motors to compare the selling price of the Pontiac Sunbird with the Chevy Cavalier. Posing as a potential customer, a representative of Teledko visited 8 Pontiac dealerships in Metro City and 6 Chevrolet dealerships and obtained quotes on comparable cars. The standard deviation for the selling prices of 8 Pontiac Sunbirds is \$350 and on the six Cavaliers, \$290. At the 0.01 significance level is there a difference in the variation in the quotes of the Pontiacs and Chevrolets?

Solution 1 – Comparing Two Populations Variances Using a Two-Tailed Test

Let the Sunbird be population 1 and the Cavalier population 2. A two-tailed test is appropriate because we are looking for a difference in the variances. We are not trying to show that one population has a larger variance than the other. The null and alternate hypotheses are:

$$H_o: \sigma_1^2 = \sigma_2^2$$
$$H_1: \sigma_1^2 \neq \sigma_2^2$$

The F distribution is the appropriate test statistic for comparing two sample variances. For a two-tailed test, the larger sample variance is placed in the numerator. The critical value of F is found by dividing the significance level in half and then referring to Appendix G and the appropriate degrees of freedom. There are $(n-1) = (8-1) = 7$ degrees of freedom in the numerator and $(n-1) = (6-1) = 5$ degrees of freedom in the denominator. From Appendix G, using the 0.01 significance level, the critical value of F is 10.5. If the ratio of the two variances exceeds 10.5, the null hypothesis is rejected and the alternate hypothesis is accepted. The computed value of the test statistic is determined by formula [12-1]:

$$F = \frac{s_1^2}{s_2^2} = \frac{(350)^2}{(290)^2} = 1.46$$

The null hypothesis is not rejected. There is no difference in the variation in the price quotes of Pontiac and Chevrolet, because the computed value of F (1.46) is less than the critical F value (10.5).

Self-Review 12.1

Check your answers against those in the ANSWER section.

Thomas Economic Forecasting, Inc. and Harmon Econometrics have the same mean error in forecasting the stock market over the last ten years. However, the standard deviation for Thomas is 30 points and 60 points for Harmon. At the 0.05 significance level can we conclude that there is more variation in the forecast given by Harmon Econometrics?

Example 2 – Comparing Two Population Variances

Tiedke's Department Store accepts three types of credit cards, MasterCard, Visa, and their own store card. The sales manager is interested in finding out whether there is a difference in the mean amounts charged by customers on the three cards. A random sample of 18 credit card purchases (rounded to the nearest dollar) revealed these credit card amounts. At the 0.05 significance level, can we conclude there is a difference in the mean amounts charged per purchase on the three cards?

Master Card	Visa	Store
$61	$85	$61
28	56	25
42	44	42
33	72	31
51	98	29
56	56	
	72	

Solution 2 – Comparing Two Population Variances

We follow the usual five-step hypothesis testing procedure.

Step 1. State the null hypothesis and the alternate hypothesis

There are three populations involved, the three credit cards. The null and alternate hypotheses are:

$H_0: \mu_1 = \mu_2 = \mu_3$

H_1: the means are not all equal

Step 2. Select the level of significance.

We have selected 0.05.

Step 3. Identify the test statistic.

For an analysis of variance problem the appropriate test statistic is F.

Step 4. Formulate the Decision Rule.

There are three "treatments" or columns. Hence, there are $(k - 1) = (3 - 1) = 2$ degrees of freedom in the numerator. There are 18 observations, therefore $n = 18$. The number of degrees of freedom in the denominator is 15, found by $(n - k) = (18 - 3)$. The critical value is found in Appendix G. Find the table for the 0.05 significance level and the column headed by 2 degrees of freedom. Then move down that column to the margin row with 15 degrees of freedom and read the value. It is 3.68.

The decision rule is: Reject the null hypothesis if the computed value of F exceeds 3.68, otherwise do not reject H_0. Shown graphically, the decision rule is shown at the right.

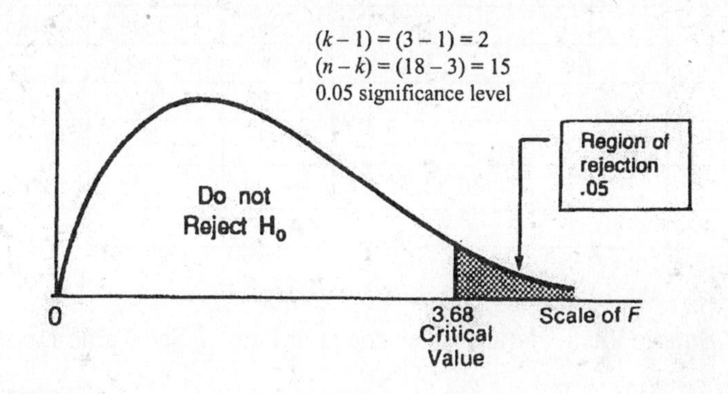

Step 5. Select the sample, perform the calculations, and make a decision.

There are two sources of variation in an analysis of variance study. These sources occur between treatments (designated SST) and within treatments (designated SSE). The sum of SST and SSE is the total amount of variation, written SS total; n refers to the number of observations in each column (treatments).

The calculations needed for SS total, SSE, and SST are as follows.

1. To calculate the **SS total** we use text formula [12–2]. $SS\,total = \Sigma(X - \bar{X}_G)^2$

   We start by determining the overall grand mean $\bar{X}_G$. The formula is: $\bar{X}_G = \dfrac{\Sigma all\ the\ X\ values}{n}$, Where: $\bar{X}_G$ is the grand mean, X is each sample observation, and n is the total number of observations.

   Using Table I, columns **A, D, & G**: $\bar{X}_G = \dfrac{\Sigma all\ the\ X\ values}{n} = \dfrac{271+483+188}{6+7+5} = \dfrac{942}{18} = 52.33$

2. Find the deviations of each observation from the grand mean. For the MC observation of \$61, the deviation is $(61 - 52.33) = 8.67$.See Table I, columns **B, E, & H**.

Table I

	A	B	C	D	E	F	G	H	I	J
	MC			Visa			Store			
	X	$(X-\bar{X}_G)$	$(X-\bar{X}_G)^2$	X	$(X-\bar{X}_G)$	$(X-\bar{X}_G)^2$	X	$(X-\bar{X}_G)$	$(X-\bar{X}_G)^2$	**Totals**
	$61	8.67	75.17	$85	32.67	1067.33	$61	8.67	75.17	
	28	–24.33	591.95	56	3.67	13.47	25	–27.33	746.93	
	42	–10.33	106.71	44	–8.33	69.39	42	–10.33	106.71	
	33	–19.33	373.65	72	19.67	386.91	31	–21.33	454.97	
	51	–1.33	1.77	98	45.67	2085.75	29	–23.33	544.29	
	56	3.67	13.47	56	3.67	13.47				
				72	19.67	386.91				
Σ	271			483			188			942
Σ			1162.72			4023.23			1928.07	7114.02
n	6			7			5			18

3. Square the deviations from the grand mean. See Table I, columns **C, F, & I.**

4. The **SS total** is found using text formula [12–2]. We summed the totals for columns **C, F, & I.**

$$SS\,total = \Sigma\left(X-\bar{X}_G\right)^2 = (1162.72+4023.23+1928.07) = 7114.02$$

5. Compute **SSE** or the sum of the squared errors. This is the sum of the squared differences between each observation and its respective treatment mean.

Using Table II, For the MasterCard observations in column **A**, the mean is: $\bar{X} = \frac{\Sigma X}{n} = \frac{271}{6} = 45.17$

In Column **B**, We subtract each observation from the mean 45.17 and then square this difference as shown in Column **C**. Then compute the sum. It is 854.84.

This process is repeated for the other two treatments or sets of observations Visa and Store in the remaining columns **D** to **I**.

The mean for Visa is $\bar{X} = \frac{\Sigma X}{n} = \frac{483}{7} = 69.00$ and for Store it is $\bar{X} = \frac{\Sigma X}{n} = \frac{188}{5} = 37.60$

Table II

	A	B	C	D	E	F	G	H	I
	MC	$\bar{X}=45.17$		Visa	$\bar{X}=69.00$		Store	$\bar{X}=37.60$	
	X	$(X-\bar{X})$	$(X-\bar{X})^2$	X	$(X-\bar{X})$	$(X-\bar{X})^2$	X	$(X-\bar{X})$	$(X-\bar{X})^2$
	$61	15.83	250.59	$85	16.00	256.00	$61	23.40	547.56
	28	–17.17	294.81	56	–13.00	169.00	25	–12.60	158.76
	42	–3.17	10.05	44	–25.00	625.00	42	4.40	19.36
	33	–12.17	148.11	72	3.00	9.00	31	–6.60	43.56
	51	5.83	33.99	98	29.00	841.00	29	–8.60	73.96
	56	10.83	117.29	56	–13.00	169.00			
				72	3.00	9.00			
Σ	271			483			188		
Σ			854.84			2078.00			843.20
n	6			7			5		

The **sum of squared errors** (SSE) is computed using text formula [12–3]. We summed the totals for columns **C, F, & I.**

$$SSE = \Sigma(X - \bar{X}_c)^2 = (854.84 + 2078.00 + 843.20) = 3776.04$$

6. Now we need to compute the **sum of squares due to treatment** (SST) using text formula [12–4]

$$SST = SS\,total - SSE = (7114.02 - 3776.04) = 3337.98$$

7. The next step is to insert these values into the ANOVA table.

Source Variation	Sum of Squares	Degrees of Freedom	Mean Squares
Treatment	SST = 3,337.98	$(k-1) = (3-1) = 2$	$\text{MST} = \frac{SST}{(k-1)} = \frac{3,337.98}{2} = 1,668.99$
Error	SSE = 3,776.04	$(n-k) = (18-3) = 15$	$\text{MSE} = \frac{SSE}{(n-k)} = \frac{3,776.04}{15} = 251.736$
Total	SS Total = 7,114.02	17	

8. The "degrees of freedom" for the treatment is $(k - 1)$, where (k) is the number of treatments. The mean square for the treatments is *SST* divided by it degrees of freedom and is written as MST.

The "degrees of freedom" for the error is $(n - k)$, where (n) is the number of observations and (k) is the number of treatments. The mean square error is SSE divided by its degrees of freedom and is written MSE.

9. Computing F using the formula:

$$F = \frac{\text{MST}}{\text{MSE}} = \frac{1,668.98}{251.73} = 6.63$$

Since the computed value of F (6.63) exceeds the critical value of 3.68, the null hypothesis is rejected at the 0.05 level and the alternate hypothesis is accepted. It is concluded that mean amounts charged by Tiedke's Department Store customers are not the same for the three credit cards.

There are many computer software packages that will perform the ANOVA calculations and output the results. MINITAB, SAS, SPSSX, and Excel are examples. The following output is from the MINITAB system. Notice that computed F is the same as determined previously.

```
ANALYSIS OF VARIANCE
SOURCE      DF       SS        MS         F        P
FACTOR       2     3338      1669      6.63    0.009
ERROR       15     3776       252
TOTAL       17     7114
                                  INDIVIDUAL 95 PCT CI'S FOR MEAN
                                  BASED ON POOLED STDEV
LEVEL        N     MEAN     STDEV  ---------+---------+---------+-------
master       6    45.17     13.08          (-------*-----)
visa         7    69.00     18.61                     (------*-----)
store        5    37.60     14.52  (--------*------)
                                   ---------+---------+---------+-------
POOLED STDEV =    15.87                    40        60        80
```

Self-Review 12.2

Check your answers against those in the ANSWER section.

The accelerating cost of electricity and gas has caused the management at Arvco Electronics to lower the heat in the work areas. The instructor conducting night classes for employees is concerned that this may have an adverse effect on the employees' test scores. Management agreed to investigate. The employees taking the basic statistics course were randomly assigned to three groups. One group was in a classroom having a temperature of 60^0, another group was placed in a room having a temperature of 70^0, and the third group was in a room having a temperature of 80^0. At the completion of the chapters on tests of hypotheses a common examination was given consisting of ten questions. The number correct for each of the 20 employees is shown in the table. At the 0.05 significance level can management conclude that there is a difference in achievement with respect to the three temperatures? (Use the table to aid in the computations).

Table I

60 Degrees			70 Degrees			80 Degrees		
X			*X*			*X*		
3			7			4		
5			6			6		
4			8			5		
3			9			7		
4			6			6		
			8			5		
			8			4		
						3		

Table II

60 Degrees			70 Degrees			80 Degrees		
X			*X*			*X*		
3			7			4		
5			6			6		
4			8			5		
3			9			7		
4			6			6		
			8			5		
			8			4		
						3		

Example 3 – Computing the Confidence Interval for the Difference in Treatment Means

In Example 1 it was concluded that there was a difference between the mean amounts charged for the three different credit cards, MasterCard, Visa, and the Tiedke's Store card. Between which credit cards is there a significant difference? Use the 0.05 level of significance.

Solution 3 – Computing the Confidence Interval for the Difference in Treatment Means

From the MINITAB Output on the page 218 note that the mean amount charged using the VISA card was \$69.00 and \$37.60 for the Department Store card. Since these means have the largest difference, let's determine if this pair of means differ significantly.

To determine if the means differ, we develop a confidence interval for the difference between the two population means. This confidence interval employs the *t* distribution and the mean square error (MSE) term. Recall that one of the assumptions for ANOVA is that the standard deviations (or variances) in the sampled populations must be the same. The MSE term is an estimate of this common variance. It is obtained from the MINITAB output. Text formula [12-5] is used:

Confidence Interval for the Difference in Treatment Means

$$(\bar{X}_1 - \bar{X}_2) \pm t\sqrt{MSE\left(\frac{1}{n_1} + \frac{1}{n_2}\right)} \quad [12-5]$$

Where:

$\bar{X}_1$ is the mean of the first treatment or sample.

$\bar{X}_2$ is the mean of the second treatment or sample.

t is obtained from the t table in Appendix F. The degree of freedom is equal to $(n - k)$.

MSE is the mean square error term, which is obtained from the ANOVA table. It is equal to SSE/$(n - k)$ and is an estimate of the common population variance.

n_1 is the number of observations in the first sample.

n_2 is the number of observations in the second sample.

If the confidence interval includes 0, there is not a difference in the treatment means. However, if both end points of the confidence interval are on the same side of 0 the pair of means differs.

$$(\bar{X}_1 - \bar{X}_2) \pm t\sqrt{MSE\left(\frac{1}{n_1} + \frac{1}{n_2}\right)}$$

$$(69.00 - 37.60) \pm 2.131\sqrt{252\left(\frac{1}{7} + \frac{1}{5}\right)}$$

$$31.40 \pm 19.81$$

$$\$11.59 \text{ to } \$51.21$$

Where:

$\bar{X}_1$ equals 69.00

$\bar{X}_2$ equals 37.60

n_1 equals 7

n_2 equals 5

t is 2.131 from Appendix F with 15 degrees of freedom and the 95 percent level of confidence.

MSE is 252, which is in the ANOVA table constructed to calculate F.

Since both end points have the same sign, positive in this case, we conclude that there is a difference in the mean amount charged on VISA and the store card.

Similarly, approximate results can be obtained directly from the MINITAB Output. In the lower right corner of the Output a confidence interval was developed for each mean. The * indicates the mean of the treatment and the parentheses ( ) symbols indicate the endpoints of the confidence interval. In comparing treatment means, if there is any common area between the two, they do not differ. If there is not any common area between the treatment means, they differ. For the credit card example, MasterCard and VISA have common area and do not differ. MasterCard and the store card do not differ, but the store and VISA do differ.

Self-Review 12.3

Check your answers against those in the ANSWER section.

In Self-Review 12.2 it was concluded that there was a difference between the mean number correct for the three different temperatures in the work area. The mean score for the 60^0 group is 3.8 and the mean score for the 70^0 group is 7.4. Determine if this pair of means differs significantly.

CHAPTER 12 ASSIGNMENT

ANALYSIS OF VARIANCE

Name ______________________ Section __________ Score ________

Part I Select the correct answer and write the appropriate letter in the space provided.

______ **1.** The analysis of variance technique is a method for
- **a.** comparing three or more means.
- **b.** comparing *F* distributions.
- **c.** measuring sampling error.
- **d.** none of the above.

______ **2.** A treatment is
- **a.** a normal population.
- **b.** the explained population.
- **c.** a source of variation.
- **d.** the amount of random error.

______ **3.** In a one-way ANOVA, *k* refers to the
- **a.** number of observations in each column.
- **b.** the number of treatments.
- **c.** the total number of observations.
- **d.** none of the above.

______ **4.** The *F* distribution is
- **a.** a continuous distribution.
- **b.** based on two sets of degrees of freedom.
- **c.** never negative.
- **d.** all of the above

______ **5.** In an ANOVA test there are 5 observations in each of three treatments. The degrees of freedom in the numerator and denominator respectively are:
- **a.** 2, 4
- **b.** 3, 15
- **c.** 3, 12
- **d.** 2, 12

______ **6.** Which of the following assumptions is **not** a requirement for ANOVA?
- **a.** dependent samples
- **b.** normal populations
- **c.** equal population variances
- **d.** independent samples

______ **7.** The mean square error term (MSE) is the
- **a.** estimate of the common population variance.
- **b.** estimate of the population means.
- **c.** estimate of the sample standard deviation.
- **d.** treatment variation.

______ **8.** In a one-way ANOVA, the null hypothesis indicates that the treatment means
- **a.** are all the same or from equal populations.
- **b.** are not from the same populations.
- **c.** in at least one pair of means are the same.
- **d.** are all different.

______ **9.** The appropriate test statistic for comparing two sample variances to find out if they came from the same or equal populations is the
- **a.** *t* distribution.
- **b.** *z* distribution.
- **c.** *F* distribution.
- **d.** binomial distribution.

______ **10.** What is the probability for an *F* of more than 6.55 with 3 degrees of freedom in the numerator and 10 in the denominator?
- **a.** 0.025
- **b.** 0.001
- **c.** 0.01
- **d.** 0.05

Part II Record your answer in the space provided. Show essential calculations.

11. The NPC, Inc. is a large mail order company that ships men's shirts all over the United States and Canada. They ship a large number of packages from their warehouse in Delta, Ohio. Their goal is to have 95 percent of the shipments delivered in 4 days. For many years they have used Brown Truck Inc., but recently there have been complaints about slow and inconsistent delivery. A sample of 10 recent shipments handled by Brown Truck showed a standard deviation in delivery time of 1.25 days. A sample of 16 shipments by Rapid Package Service showed a standard deviation in their delivery time of 0.45 days. At the 0.05 significance level is there more variation in the Brown Truck delivery time?

a. State the null and alternate hypotheses.

H_0: ______________________________

H_1: ______________________________

b. State the decision rule.

____

c. Compute the value of the test statistic.

c.

d. What is your decision regarding the null hypothesis? Interpret the result.

____

12. Jim Ray, an avid golfer, keeps records on his scores for 18 holes of golf. When the temperature is above 65 degrees, the standard deviation of his scores is 5.75 for 25 rounds. When the temperature is below 65 degrees, the standard deviation of his scores is 7.35 for a sample of 21 rounds. At the 0.05 significance level, is there more variation in his scores when the temperature is below 65 degrees?

a. State the null and alternate hypotheses.

H_0: ______________________________

H_1: ______________________________

b. State the decision rule.

____

c. Compute the value of the test statistic.

c.

d. What is your decision regarding the null hypothesis? Interpret the result.

____

13. The County Executive for Monroe County is concerned about the response time for the three fire companies in the county. Samples of the response times (in minutes) for each company follow. At the 0.05 significance level is there a difference in the mean response time?

Youngsville	Northeast	Corry
2.2	2.3	0.9
1.2	1.5	0.8
1.9	1.2	1.1
3.1	1.4	1.2
1.8	2.2	0.7
1.5		

a. State the null and alternate hypotheses.

H_0: ____

H_1: ____

b. State the decision rule.

____

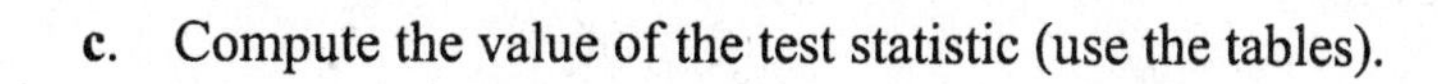

c. Compute the value of the test statistic (use the tables).

c.

d. What is your decision regarding the null hypothesis? Interpret the result.

CHAPTER 13
LINEAR REGRESSION AND CORRELATION

Chapter Goals

After completing this chapter, you will be able to:

1. Understand and interpret the terms ***dependent variable*** and ***independent variable***.
2. Calculate and interpret the ***coefficient of correlation***, the ***coefficient of determination***, and the ***standard error of estimate***.
3. Conduct a test of hypothesis to determine whether the coefficient of correlation in the population is zero.
4. Calculate the least squares regression line.
5. Construct and interpret confidence intervals and prediction intervals for the dependent variable.
6. Set up and interpret an ANOVA table.

Introduction

We studied hypothesis testing concerning means and proportions where only a single feature of the sampled item was considered. For example, based on sample evidence we concluded that the beginning annual mean salary for accounting graduates is $26,000. With this chapter we begin our study of the relationship between two variables. We may want to determine if there is a relationship between the number of years of company service and the income of executives. Or we may want to explore the relationship between crime in the inner city and the unemployment rate.

What is Correlation Analysis?

To study the relationship between two variables we use two techniques: ***correlation analysis*** and ***regression analysis.***

> ***Correlation analysis***: A group of techniques to measure the association between two variables.

The purpose of correlation analysis is to find the relationship between two variables. One way of looking at the relationship between two variables is to portray the information in a ***scatter diagram*** as described in Chapter 4.

> ***Scatter diagram***: A graph in which paired data values are plotted on an *X,Y Axis*.

The values of the ***independent variable*** (x) are portrayed on the horizontal axis (X-axis) and the ***dependent variable*** (y) along the vertical axis (Y-axis).

> ***Dependent variable***: The variable that is being predicted or estimated.

> ***Independent variable***: A variable that provides the basis for estimation. It is the predictor variable.

Note in Figure A that as the length of service increases so does income. In Figure B, as employment rises, the crime rate in the inner city declines.

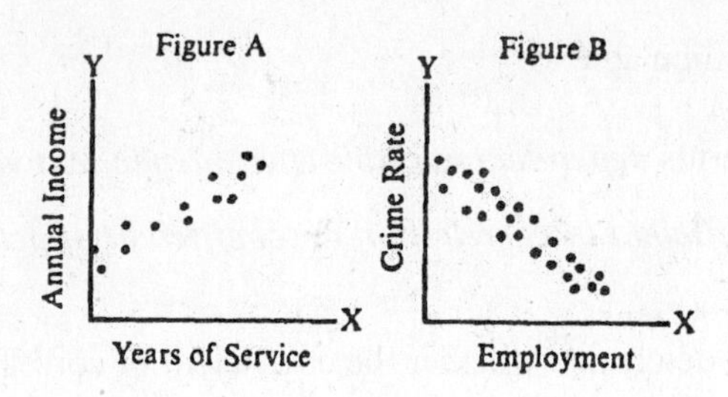

The Coefficient of Correlation

A measure of the linear (straight-line) strength of the relationship between two sets of interval-scaled or ratio-scaled variables is given by the ***coefficient of correlation***.

> ***Coefficient of correlation***: A measure of the strength of the linear relationship between two sets of interval-scaled or ratio-scaled variables.

The coefficient of correlation is also called **Pearson's product moment correlation coefficient** or **Pearson's** r after its founder Karl Pearson. The sample correlation coefficient is designated by the lower case r. Its value may range from –1.00 to +1.00 inclusive. A value of –1.00 indicates perfect negative correlation. A value of +1.00 indicates perfect positive correlation. A correlation coefficient of 0.0 indicates there is no relationship between the two variables under consideration. This information is summarized in the charts that follow.

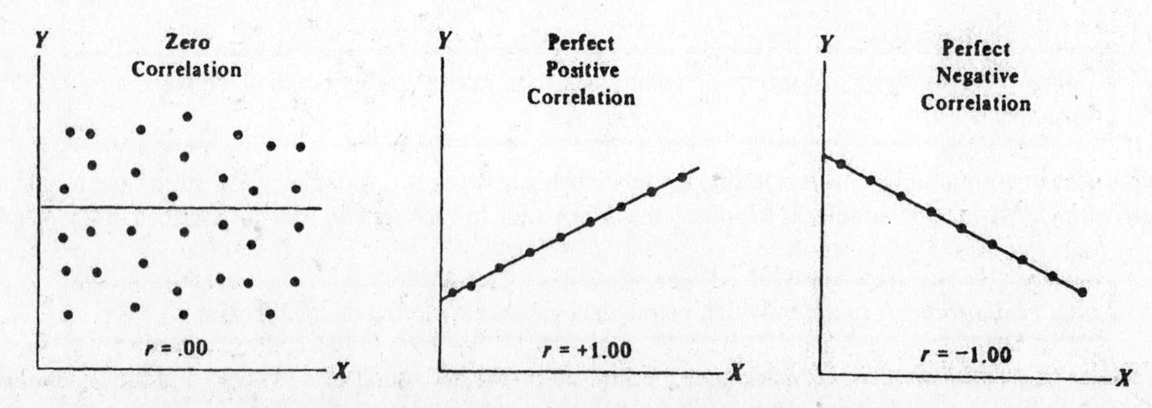

The coefficient of correlation requires that both variables be at least of interval scale.

The degree of strength of the relationship is not related to the sign (direction – or +) of the coefficient of correlation. For example, an r value of –0.60 represents the same degree of correlation as +0.60. An r of –0.70

represents a stronger degree of correlation than 0.40. An r of -0.90 represents a strong negative correlation and $+0.15$ a weak positive correlation.

The coefficient of correlation is computed by text formula [13-1].

Correlation Coefficient $$r = \frac{\Sigma(X - \bar{X})(Y - \bar{Y})}{(n-1)(s_x s_y)} \quad [13-1]$$

Where:

r	is the coefficient of correlation.
n	is the number of paired observations.
$\bar{X}$	is the mean for the X variable.
$\bar{Y}$	is the mean for the Y variable.
s_x	is the standard deviation of the X values.
s_y	is the standard deviation of the Y values.

Testing the Significance of the Correlation Coefficient

A test of significance for the coefficient of correlation may be used to determine if the computed r could have occurred in a population in which the two variables are not related. To put it in the form of a question: Is the correlation in the population zero?

For a two-tailed test the null hypothesis and the alternate hypothesis are written as follows:

H_0: $\rho = 0$ (The correlation in the population is zero)

H_1: $\rho \neq 0$ (The correlation in the population is different from zero)

The Greek lower case rho, ρ, represents the correlation in the population. The null hypothesis is that there is no correlation in the population, and the alternate that there is a correlation.

From the way H_1 is stated, we know that the test is two tailed. The alternate hypothesis can also be set up as a one-tailed test. It could read, "The correlation coefficient is greater than zero."

The test statistic follows the t distribution with $(n - 2)$ degrees of freedom. Text formula [13-2] is used.

***t* Test for the Coefficient of Correlation** $$t = \frac{r\sqrt{n-2}}{\sqrt{1-r^2}} \quad [13-2]$$

Regression Analysis

We have an equation that shows the linear (straight line) relationship between two variables. The equation is used to estimate Y based on X and is referred to as the ***regression equation***.

Regression Equation: An equation that defines the linear relationship between two variables.

The linear relationship between two variables is given by the general form of the regression equation. Text formula [13-3] is used.

General Form of Linear Regression Equation	$Y' = a + bX$	[13−3]

Where:

Y' (read Y prime) is the value of the Y variable for a selected X value.

a is the Y intercept. It is the value of Y when $X = 0$.

b is the slope of the line. It measures the change in Y' for each unit change in X. It will always have the same sign as the coefficient of correlation.

X is any value of the independent variable that is selected.

The value of ***a*** is the ***Y* intercept** and is computed using formula [13-5].

Y-Intercept	$a = \bar{Y} - b\bar{X}$	[13−5]

where:

$\bar{Y}$ is the mean of the $Y-$ values (the dependent variable).

$\bar{X}$ is the mean of the $X-$ values (the independent variable).

The value of ***b*** is the slope of the **regression line** and is computed using formula [13-4].

Slope of the Regression Line	$b = r\left(\frac{s_y}{s_x}\right)$	[13-4]

Where:

r is the correlation coefficient.

s_x is the standard deviation of the X values (the dependent variable).

s_y is the standard deviation of the Y values (the independent variable).

How do we get these values? They are developed mathematically using the ***least squares principle***.

Least squares principle: Determining a regression equation by minimizing the sum of the squares of the vertical distances between the actual Y values and the predicted values of Y'.

Suppose the least squares principle was used to develop an equation expressing the relationship between annual salary and years of work experience. The equation is:

$$Y' = a + bX$$
$$= 20{,}000 + 500X \text{ (in dollars)}$$

In the example, annual income is the dependent variable, Y, and is being predicted on the basis of the employee's years of work experience, X, the independent variable. The value of 500, which is b, means that for each additional year of work experience the employee's salary increases by $500. Thus, we would expect an employee with 40 years of work experience to earn $5,000 more than one with 30 years of work experience.

What does the 20,000 dollars represent? It is the value for Y' when $X = 0$. Recall that this is the point where the line intersects the Y-axis. The values of a and b in the regression equation are usually referred to as the regression coefficients.

The Standard Error of Estimate

Rarely does the predicted value of Y' agree exactly with the actual Y value. That is, we expect some prediction error. One measure of this error is called the ***standard error of estimate***. It is written $s_{y.x}$.

> ***Standard error of estimate***: A measure of the scatter, or dispersion, of the observed values around the line of regression.

A small standard error of estimate indicates that the independent variable is a good predictor of the dependent variable.

The standard error, as it is often called, is similar to the standard deviation described in Chapter 4. Recall that the standard deviation was computed by squaring the difference between the actual value and the mean. This squaring was performed for all n observations. For the standard error of estimate, the difference between the predicted value Y' and the actual value of Y is obtained and that difference squared and summed over all n observations. The text formula [13-6] is:

Standard Error of Estimate

$$s_{y \cdot x} = \sqrt{\frac{\Sigma(Y - Y')^2}{n-2}} \quad [13-6]$$

Assumptions Underlying Linear Regression

Linear regression is based on these four assumptions:

1. For each value of X, there is a group of Y values. These Y values follow the normal distribution.
2. The means of these normal distributions of Y values all lie on the line of regression.
3. The standard deviations of these normal distributions are all the same. The best estimate we have of this common standard deviation is the standard error of estimate $(s_{y \cdot x})$.
4. The X values are statistically independent. This means in selecting a sample, a particular X does not depend on any other value of X. This assumption is particularly important when data are collected over a period of time. In such situations, the errors for a particular time period are often correlated with those of other time periods.

Confidence and Prediction Intervals

The standard error is also used to set confidence intervals for the predicted value of Y'. When the sample size is large and the scatter about the regression line is approximately normally distributed, then the following relationships can be expected:

$Y' \pm 1s_{y \cdot x}$	encompasses about 68% of the observed values.
$Y' \pm 2s_{y \cdot x}$	encompasses about 95% of the observed values.
$Y' \pm 3s_{y \cdot x}$	encompasses virtually all of the observed values.

Two types of ***confidence intervals*** may be set. The first for the mean value of Y' for a given value of X and the other, called a ***prediction interval,*** for an individual value of Y' for a given value of X. To explain the difference between the mean predicted value and the individual prediction, suppose we are predicting the salary of management personnel who are 40 years old. In this case we are predicting the mean salary of all management personnel age 40. However, if we want to predict the salary of a particular manager who is 40, then we are making a prediction about a particular individual.

The formula for the confidence interval for the mean value of Y for a given X is:

Confidence Interval for the mean of Y, given X

$$Y' \pm t(s_{y \cdot x})\sqrt{\frac{1}{n}+\frac{(X-\bar{X})^2}{\Sigma(X-\bar{X})^2}} \qquad [13-7]$$

Where:

Y' is the predicted value for a selected value of X.

X is any selected value of the independent variable.

$\bar{X}$ is the mean of the independent variable X, found by $\Sigma X \div n$.

n is the sample size or number of observations.

t is the value of the Student t distribution from Appendix F, with $(n - 2)$ degrees of freedom and the given level of significance for a two-tailed test.

$s_{y.x}$ is the standard error of estimate.

The formula is modified slightly for a prediction interval. The number 1 (one) is placed under the radical and the formula becomes:

Prediction Interval for Y, given X

$$Y' \pm t(s_{y \cdot x})\sqrt{1+\frac{1}{n}+\frac{(X-\bar{X})^2}{\Sigma(X-\bar{X})^2}} \qquad [13-8]$$

The Relationship Among Various Measures of Association

The standard error of estimate measures how closely the actual values of Y are to the predicted values of Y'. When the values are close together the standard error is "small." When they are spread out, the standard error will be large. In the calculation of the standard error, the key term is: $\Sigma(Y - Y')^2$

When this term is small, the standard error is also small.

Recall that the coefficient of correlation measured the strength of the association between two variables. When the points on a scatter diagram are close to a straight line, the correlation coefficient tends to be "large." Thus the standard error and the coefficient of correlation reflect the same information but use a different scale to report it. The standard error is in the same units as the dependent variable. The correlation coefficient has a range of −1.00 to +1.00.

The ***coefficient of determination*** also reports the strength of the association.

> ***Coefficient of determination***: The proportion of the total variation in the dependent variable *Y* that is explained or accounted for, by the variation in the independent variable *X*.

It is the square of the correlation coefficient and has a range 0.00 to 1.00.

A convenient means of showing the relationships among these measures is an ANOVA Table. This is similar to the table developed in the previous chapter. The total variation $\Sigma(Y-\bar{Y})^2$ is divided into two components.

1. The component explained by the regression.
2. The error or unexplained variation.

These two categories are identified in the source column of the following ANOVA table. The column headed *DF* refers to the degrees of freedom associated with each category. The total degrees of freedom is $(n-1)$. The degrees of freedom in the regression is 1 because there is one independent variable. The degrees of freedom associated with the error term is $(n-2)$. The term SS, located in the middle of the table, refers to the variation. These terms are computed as follows:

$$\text{Total variation} = \text{SS total} = \Sigma(Y-\bar{Y})^2$$

$$\text{Error variation} = \text{SSE} = \Sigma(Y-Y')^2$$

$$\text{Regression} = \text{SSR} = \Sigma(Y'-\bar{Y})^2$$

The format for the ANOVA table is:

Source	**DF**	**SS**	**MS**
Regression	1	SSR	SSR/1
Error	n − 2	SSE	SSE/(n − 2)
Total	$n-1$	SS total*	

*SS total = SSR + SSE

The coefficient of determination, r^2 can be computed directly from the ANOVA table.

Coefficient of Determination

$$r^2 = \frac{SSR}{SS\text{ total}} = 1 - \frac{SSE}{SS\text{ total}} \qquad [13-10]$$

Note that as SSE decreases r^2 increases. The coefficient of correlation is the square root of this value. Hence, both of these values are related to SSE. The standard error of estimate is obtained using the following equation.

Standard Error of Estimate	$s_{y \cdot x} = \sqrt{\frac{SSE}{n-2}}$	[13–11]

Note again the role played by the SSE term. A small value of SSE will result in a small standard error of estimate.

Glossary

***Coefficient of correlation*:** A measure of the strength of the linear relationship between two sets of interval-scaled or ratio-scaled variables.

***Coefficient of determination*:** The proportion of the total variation in the dependent variable Y that is explained, or accounted for, by the variation in the independent variable X.

***Correlation analysis*:** A group of techniques used to measure the strength of the association between two variables.

***Dependent variable*:** The variable that is being predicted or estimated.

***Independent variable*:** A variable that provides the basis for estimation. It is the predictor variable.

***Least squares principle*:** Determining a regression equation by minimizing the sum of the squares of the vertical distances between the actual Y values and the predicted values of Y'.

***Regression equation*:** An equation that defines the linear relationship between two variables.

***Scatter diagram*:** A graph in which paired data values are plotted on an *X,Y Axis*.

***Standard error of estimate*:** A measure of the scatter, or dispersion, of the observed values around the line of regression.

Chapter Examples

Example 1 – Computing the Coefficient of Correlation

It is believed that the annual repair cost for a commercial dishwasher in a restaurant is related to its age. A sample of 10 dishwashers revealed the results in the table at the right.

Repair Cost (in dollars) Y	Age (in years) X
$72	2
99	3
65	1
138	7
170	6
140	8
114	4
83	1
101	2
110	5

a. Plot these data in a scatter diagram. Does it appear there is a relationship between repair cost and age?
b. Compute the coefficient of correlation.
c. Determine at the 0.05 significance level whether the correlation in the population is greater than zero.

Solution 1 – Computing the Coefficient of Correlation

a. The repair cost is the dependent variable and is plotted along the Y-axis. Age is the independent variable and is plotted along the X-axis. To plot the first point, move horizontally on the X-axis to 2 and then go vertically to 72 on the Y-axis and place a dot. This procedure is continued until all paired data are plotted. Note that it appears there is a positive relationship between the two variables. That is, as X, the age of the dishwasher increases, so does the repair cost. But, the relationship is not perfect as evidenced by the scatter of dots.

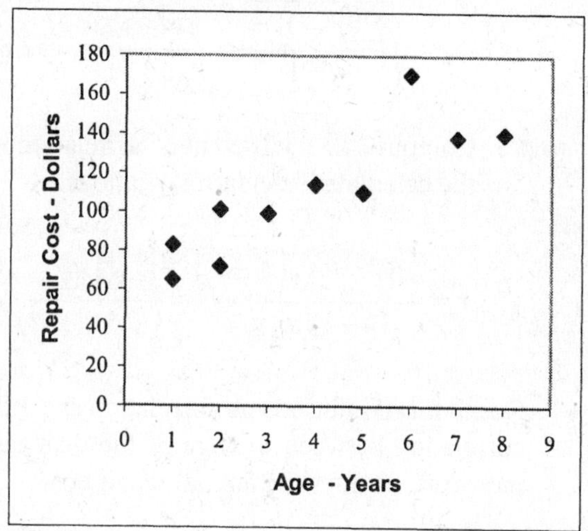

b. The degree of association between age and repair cost is measured by the coefficient of correlation. It is computed by formula [13-1]. Note that we also need to compute the mean and the standard deviation using formula [3-2] and [3-11].

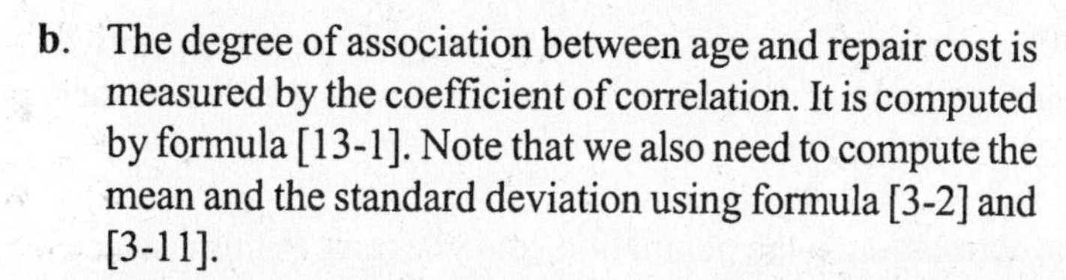

$$r = \frac{\Sigma(X-\bar{X})(Y-\bar{Y})}{(n-1)(s_x s_y)} \quad [13-1] \qquad \bar{X} = \frac{\Sigma X}{n} \quad [3-2] \qquad s = \sqrt{\frac{\Sigma(X-\bar{X})^2}{n-1}} \quad [3-11]$$

The calculations in the table below are needed to compute the various totals.

	A	B	C	D	E	F	G
	Repair Cost			Age			
	Y	$(Y-\bar{Y})$	$(Y-\bar{Y})^2$	X	$(X-\bar{X})$	$(X-\bar{X})^2$	$(X-\bar{X})(Y-\bar{Y})$
	72	-37.2	1,383.84	2	-1.9	3.61	70.68
	99	-10.2	104.04	3	-0.9	0.81	9.18
	65	-44.2	1,953.64	1	-2.9	8.41	128.18
	138	28.8	829.44	7	3.1	9.61	89.28
	170	60.8	3,696.64	6	2.1	4.41	127.68
	140	30.8	948.64	8	4.1	16.81	126.28
	114	4.8	23.04	4	0.1	0.01	0.48
	83	-26.2	686.44	1	-2.9	8.41	75.98
	101	-8.2	67.24	2	-1.9	3.61	15.58
	110	0.8	0.64	5	1.1	1.21	0.88
n	10			10			
Σ	1092	0.00	9,693.60	39	0.00	56.9	644.20

Step 1. Compute the means using sums in Column **A** and **D**:

$$\bar{Y} = \frac{\Sigma Y}{n} = \frac{1092}{10} = 109.2 \qquad \bar{X} = \frac{\Sigma X}{n} = \frac{39}{10} = 3.9$$

Step 2. Compute the standard deviations using the sums in Column **C** and **F**:

$$s_y = \sqrt{\frac{\Sigma(Y-\bar{Y})^2}{n-1}} = \sqrt{\frac{9{,}693.60}{10-1}} = 32.8 \qquad s_x = \sqrt{\frac{\Sigma(X-\bar{X})^2}{n-1}} = \sqrt{\frac{56.9}{10-1}} = 2.51$$

Step 3. Compute the coefficient of correlation ***r*** using the formula, the sum from Column **G** in the table, and the calculated standard deviations.

$$r = \frac{\Sigma(X-\bar{X})(Y-\bar{Y})}{(n-1)(s_x s_y)} = \frac{644.2}{9(2.51)(32.8)} = \frac{644.2}{740.952} = 0.869 = 0.87$$

Recall that 0 indicates no correlation and 1.00 perfect correlation. The *r* of 0.87 suggests a strong positive correlation between the age of the dishwasher and annual repair costs. As the age of the dishwasher increases, so does the annual repair cost.

c. A test of hypothesis is used to determine if the correlation in the population could be zero. In this instance, suppose we want to show that there is a positive association between the variables. Recall that the Greek letter ρ refers to the correlation in the population. The null and alternate hypotheses are written as follows:

$$H_0: \rho \leq 0$$
$$H_1: \rho > 0$$

If the null hypothesis is not rejected, it indicates that the correlation in the population could be zero. If the null hypothesis is rejected, the alternate is accepted. This indicates there is correlation in the population between the two variables and it is positive.

The test statistic follows the Student's *t* distribution with (n – 2) degrees of freedom. The alternate hypothesis given above specifies a one-tailed test in the positive direction. There are 8 degrees of freedom, found by (n – 2) = (10 – 2). The critical value for a one-tailed test using the 0.05 significance level is 1.860 (Appendix F). The decision rule is to reject the null hypothesis if the computed value of *t* exceeds 1.860. The computed value of *t* is 4.99, found by using formula [13–2.]

$$t = \frac{r\sqrt{n-2}}{\sqrt{1-r^2}} = \frac{0.87\sqrt{10-2}}{\sqrt{1-(0.87)^2}} = \frac{2.4607}{0.493} = 4.99$$

Since the computed value (4.99) exceeds the critical value of *t*, namely 1.860, the null hypothesis is rejected and the alternate accepted. It is concluded that there is a positive association between the age of the dishwasher and the annual repair cost. The *p*-value is less than 0.005.

Self-Review 13.1

Check your answers against those in the ANSWER section.
A major oil company is studying the relationship between the daily traffic count and the number of gallons of gasoline pumped at company stations. A sample of eight company owned stations is selected and the following information obtained:

a. Develop a scatter diagram with the amount of gasoline pumped as the dependent variable.
b. Compute the coefficient of correlation
c. Compute the coefficient of determination.
d. Interpret the meaning of the coefficient of determination.
e. Test to determine whether the correlation in the population is zero, versus the alternate hypothesis that the correlation is greater than zero. Use the 0.05 significance level.

Location	Total Gallons Pumped (000)	Traffic count of vehicles (000)
West St.	120	4
Willouhby St.	180	6
Mallard Rd.	140	5
Pheasant Rd.	150	5
I-75	210	8
Kinzua Rd.	100	3
Front St.	90	3
Indiana Ave.	80	2

Example 2 – Using the Least Squares Principle to Determine the Regression Equation

In Example 1 we examined the relationship between the annual repair cost of the dishwasher and its age. The correlation between the two variables was 0.867, which we considered to be a strong relationship. We conducted a test of hypothesis and concluded that the relationship between the two variables in the population was greater than zero. The same sample data is repeated in the table.

Repair Cost (in dollars)	Age (in years)
72	2
99	3
65	1
138	7
170	6
140	8
114	4
83	1
101	2
110	5

a. Use the least squares principle to determine the regression equation.

b. Compute the standard error of estimate.

c. Develop a 95 percent confidence interval for the mean repair cost for all 4-year-old dishwashers.

d. Develop a 95 percent prediction interval for the repair cost for a 4-year-old dishwasher.

Solution 2 – Using the Least Squares Principle to Determine the Regression Equation

a. The first step is to compute the regression equation using formula [13-4] for b, formula [13-5] for a. In Example 1, we computed: $r = 0.87$, $s_y = 32.8$, and $s_x = 2.51$. Thus, using formula [13-4]:

$$b = r\left(\frac{s_y}{s_x}\right) = 0.87\left(\frac{32.8}{2.51}\right) = 0.87(13.07) = 11.37$$

In Example 1, we computed: $\bar{Y} = 109.2$ and $\bar{X} = 3.9$. Thus, using formula [13-5]:

$$\begin{aligned} a &= \bar{Y} - b\bar{X} \\ &= 109.2 - [(11.37)(3.9)] \\ &= 109.2 - 44.343 = 64.857 \end{aligned}$$

Thus, the regression equation is:

$$Y' = a + bX = 64.857 + 11.37X \text{ (in dollars)}$$

Interpreting, repair costs can be expected to increase \$11.37 a year on the average. Stated differently, the repair cost of a 4-year-old dishwasher can be expected to cost \$11.37 more a year than a 3-year-old dishwasher.

b. The standard error of estimate is a measure of the dispersion about the regression line. It is similar to the standard deviation in that it uses squared differences. The differences between the value of Y' and Y are squared and summed over all n observations and then divided by $(n-2)$. The standard error is the positive square root of this value. A small value indicates a close association between the dependent and independent variable. The standard error is measured in the same units as the dependent variable. The symbol for the standard error of estimate is $s_{y \cdot x}$.

The standard error is computed using formula [13-6] and the table shown.

$$s_{y \cdot x} = \sqrt{\frac{\Sigma(Y - Y')^2}{n-2}} = \sqrt{\frac{2400.348}{10-2}}$$
$$= \sqrt{300.0435} = 17.32$$

X	bX	Y'	Y	$(Y-Y')$	$(Y-Y')^2$
2	22.74	87.597	72	-15.597	243.26641
3	34.11	98.967	99	0.033	0.001089
1	11.37	76.227	65	-11.227	126.04553
7	79.59	144.447	138	-6.447	41.563809
6	68.22	133.077	170	36.923	1363.3079
8	90.96	155.817	140	-15.817	250.17749
4	45.48	110.337	114	3.663	13.417569
1	11.37	76.227	83	6.773	45.873529
2	22.74	87.597	101	13.403	179.64041
5	56.85	121.707	110	-11.707	137.05385
				SUM	2400.348

c. The regression equation is used to estimate the repair cost of a 4-year-old dishwasher. The value of 4 is inserted for X in the equation.

$$Y' = 64.857 + 11.37X$$
$$= 64.857 + 11.37(4)$$
$$= 110.337 = 110.34 \text{ (in dollars)}$$

Thus the expected repair cost for a 4-year-old dishwasher is \$110.34.

Formula [13-7] is used if we want to develop a 95 percent confidence interval of the repair cost for all four-year-old dishwashers.

$$Y' \pm t(s_{y \cdot x})\sqrt{\frac{1}{n} + \frac{(X - \bar{X})^2}{\Sigma(X - \bar{X})^2}}$$

The necessary information for the formula is:

Y' is 110.34 as previously computed.
t is 2.306. There are $(n-2)$ degrees of freedom, or $(n-2) = (10-2) = 8$. From Appendix F, using a 95% confidence level, move down the column to 8 *df* and read the value of t.
n is 10. It is the sample size.
$s_{y \cdot x}$ is \$17.32, as computed in an earlier section of this Example.
X is 4, the age of the dishwasher.
$\bar{X}$ is the mean age of the sampled dishwashers. It is 3.9, found by $\bar{X} = 39/10$.
$\Sigma(X - \bar{X})^2$ is 56.9, found from the earlier computations.

Solving for the 95 percent confidence interval:

$$Y' \pm t(s_{y\cdot x})\sqrt{\frac{1}{n}+\frac{(X-\bar{X})^2}{\Sigma(X-\bar{X})^2}} = \$110.34 \pm 2.306(\$17.32)\sqrt{\frac{1}{10}+\frac{(4-3.9)^2}{56.9}}$$
$$= \$110.34 \pm \$12.64$$
$$= \$97.70 \text{ to } \$122.98$$

The 95 percent confidence interval for the mean amount spent on repairs to a 4-year-old dishwasher is between $97.70 and $122.98. About 95 percent of the similarly constructed intervals would include the population value.

d. Recall the 4-year-old dishwasher. The 95 percent prediction interval for the repair costs is computed as follows using formula [13-8].

$$Y' \pm t(s_{y\cdot x})\sqrt{1+\frac{1}{n}+\frac{(X-\bar{X})^2}{\Sigma(X-\bar{X})^2}} = \$110.34 \pm 2.306(\$17.32)\sqrt{1+\frac{1}{10}+\frac{(4-3.9)^2}{56.9}}$$
$$= \$110.34 \pm \$41.92$$
$$= \$68.42 \text{ to } \$152.26$$

Interpreting we would conclude that between $68.42 and $152.26 would be spent on repairs this year for a 4-year old dishwasher. About 95 percent of the similarly constructed intervals would include the population value.

Self-Review 13.2

Check your answers against those in the ANSWER section.

In Exercise 13.1 we studied the relationship between the gasoline pumped in thousands of gallons, and the traffic count at eight company owned stations. The data are repeated at the right.

a. Compute the regression equation.
b. Compute the standard error of estimate.
c. Develop a 95 percent confidence interval for the mean amount of gasoline pumped for all stations where the traffic count is 4.
d. Develop a 95 percent prediction interval for the amount of gasoline pumped at the station at Dowling Rd. and I-60, which has a count of 4 (actually 400 cars).

Location	Total Gallons of Gas Pumped (000)	Traffic count (hundreds of cars)
West St.	120	4
Willouhby St.	180	6
Mallard Rd.	140	5
Pheasant Rd.	150	5
I-75	210	8
Kinzua Rd.	100	3
Front St.	90	3
Indiana Ave.	80	2

Example 3 – Compute the Standard Error of Estimate

Use the information from Example 1 to:

a. Develop an ANOVA Table.

b. Compute the coefficient of determination.

c. Compute the coefficient of correlation.

d. Compute the standard error of estimate.

Solution 3 – Compute the Standard Error of Estimate

a. The MINITAB System was used to develop the following output.

Analysis of Variance

SOURCE	DF	SS	MS
Regression	1	7,293.4	7,293.4
Error	8	2,400.2	300.0
Total	9	9,693.6	

b. The coefficient of determination is computed using formula [13-10] as follows.

$$r^2 = \frac{SSR}{SS\ Total} = \frac{7293.4}{9693.6} = 0.752 \text{ OR } r^2 = \left(1 - \frac{SSE}{SS\ Total}\right) = \left(1 - \frac{2{,}400.2}{9{,}693.6}\right) = (1 - 0.248) = 0.752$$

c. The correlation coefficient is 0.867, found by taking the square root of 0.752. These two coefficients (0.867 and 0.752) are the same as computed earlier.

d. The standard error of estimate is computed using formula [13-11] as follows:

$$s_{y \cdot x} = \sqrt{\frac{SSE}{n-2}} = \sqrt{\frac{2{,}400.2}{10-2}} = 17.32$$

Note again the role played by the SSE term. A small value of SSE will result in a small standard error of estimate.

Self-Review 13.3

Check your answers against those in the ANSWER section.

Refer to Exercise 1, regarding the relationship between the amount of gasoline pumped and the traffic count. The following output was obtained from MINITAB.

Analysis of Variance

SOURCE	DF	SS	MS
Regression	1	14,078	14,078
Error	6	310	51.67
Total	7	14,388	

a. Compute the coefficient of determination.
b. Compute the coefficient of correlation.
c. Compute the standard error of estimate.

CHAPTER 13 ASSIGNMENT

LINEAR REGRESSION AND CORRELATION

Name ______________________________ Section __________ Score______

Part I Select the correct answer and write the appropriate letter in the space provided.

______ 1. Which of the following statements is **not** correct regarding the coefficient of correlation.
- **a.** It can range from –1 to 1.
- **b.** Its square is the coefficient of determination.
- **c.** It measures the percent of variation explained.
- **d.** It is a measure of the association between two variables.

______ 2. The coefficient of determination
- **a.** is usually written as r^2.
- **b.** cannot be negative.
- **c.** is the square of the coefficient of correlation.
- **d.** all of the above.

______ 3. The coefficient of correlation was computed to be –0.60. This means
- **a.** the coefficient of determination is $\sqrt{0.6}$.
- **b.** as X increase Y decreases.
- **c.** X and Y are both 0.
- **d.** as X decreases Y decreases.

______ 4. Which of the following is a stronger correlation than –0.54?
- **a.** 0.67 **b.** 0.45
- **c.** 0.0 **d.** –0.45

______ 5. A regression equation is used to
- **a.** measure the association between two variables.
- **b.** estimate the value of the dependent variable based on the independent variable.
- **c.** estimate the value of the independent variable based on the dependent variable.
- **d.** estimate the coefficient of determination.

______ 6. A regression equation was computed to be $Y' = 35 + 6X$. The value of the 35 indicates that
- **a.** the regression line crosses the Y-axis at 35.
- **b.** the coefficient of correlation is 35.
- **c.** the coefficient of determination is 35.
- **d.** an increase of one unit in X will result in an increase of 35 in Y.

______ 7. The standard error of estimate
- **a.** is a measure of the variation around the regression line.
- **b.** cannot be negative.
- **c.** is in the same units as the dependent variable.
- **d.** all of the above.

_____ 8. The variable plotted on the horizontal or *X*-axis in a scatter diagram is called the

a. scatter variable.
b. independent variable.
c. dependent variable.
d. correlation variable.

_____ 9. The least squares principle means that

a. $\Sigma(Y - Y')^2 = 0$.
b. $\Sigma(Y - \overline{Y})^2$ is maximized.
c. $\Sigma(Y - Y')^2$ is minimized.
d. $\Sigma(Y - \overline{Y})^2$ is minimized.

_____ 10. If all the points are on the regression line, then

a. the value of *b* is 0.
b. the value of *a* is 0.
c. the correlation coefficient is 0.
d. the standard error of estimate is 0.

Part II Record your answers in the space provided. Show all essential work.

11. The correlation between the number of police on the street and the number of crimes committed, for a sample of 15 comparable sized cities, is 0.45. At the 0.05 significance level is there a positive association in the population between the two variables?

a. State the null and alternate hypotheses.

H_0: ______________________________

H_1: ______________________________

b. State the decision rule.

____

c. Compute the value of the test statistic.

c.

d. What is your decision regarding the null hypothesis? Interpret the result.

____

12. A study is conducted concerning automobile speeds and fuel consumption rates. The following data is collected:

a. Plot these data in a scatter diagram.

Speed	MPG
44	22
51	26
48	21
60	28
66	33
61	32

b. Compute the coefficient of correlation.

	A	B	C	D	E	F	G
	MPG			**Speed**			
	Y			*X*			
	22			44			
	26			51			
	21			48			
	28			60			
	33			66			
	32			61			
n							
Total							

b.

c. Determine at the 0.05 significance level whether the correlation in the population is greater than zero.

c.

13. Tem Rousos, president of Rousos Ford, believes there is a relationship between the number of new cars sold and the number of sales people on duty. To investigate he selects a sample of eight weeks and determines the number of new cars sold and the number of sales people on duty for that week.

Week	Sales staff	Cars sold
1	5	53
2	5	47
3	7	48
4	4	50
5	10	58
6	12	62
7	3	45
8	11	60

a. Plot these data in a scatter diagram.

b. Determine the coefficient of correlation.

	A	B	C	D	E	F	G
				Number			
	Number of Cars Sold			of Sales Staff			
	Y			*X*			
	53			5			
	47			5			
	48			7			
	50			4			
	58			10			
	62			12			
	45			3			
	60			11			
n							
Total							

b.

c. Determine the coefficient of determination. Comment on the strength of the association between the two variables.

c.

___

___

d. Determine the regression equation.

d.

e. Interpret the regression equation. Where does the equation cross the *Y*-axis? How many additional cars can the dealer expect to sell for each additional salesperson employed?

___

___

f. Determine the standard error of estimate.

X			Y		
5			53		
5			47		
7			48		
4			50		
10			58		
12			62		
3			45		
11			60		

f.

g. Develop a 95 percent confidence interval for all the mean car sales for weeks when the sales staff is at 10.

g.

h. In checking the work schedules for next week, Tem finds there are 10 people scheduled. Develop a 95 percent prediction interval for the number of cars sold next week.

h.

CHAPTER 14
MULTIPLE REGRESSION AND CORRELATION ANALYSIS

Chapter Goals

After completing this chapter, you will be able to:

1. Describe the relationship between several independent variables and a dependent variable using a multiple regression equation.
2. Compute and interpret the multiple standard error of estimate and the coefficient of determination.
3. Interpret a correlation matrix.
4. Setup and interpret an ANOVA table.
5. Conduct a test of hypothesis to determine whether regression coefficients differ from zero.
6. Conduct a test of hypothesis on each of the regression coefficients.

Introduction

In the last chapter we began our study of regression and correlation analysis. However, the methods presented considered only the relationship between one dependent variable and one independent variable. The possible effect of other independent variables was ignored. For example, we described how the repair cost of a commercial dishwasher was related to the age of the dishwasher. Are there other factors that affect the repair cost? Does the size of the dishwasher or the temperature of the water affect the repair cost? When several independent variables are used to estimate the value of the dependent variable it is called ***multiple regression***.

> ***Multiple Regression***: A set of techniques used to analyze the relationship between two or more independent variables and a dependent variable.

Multiple Regression Analysis

Recall that for one independent variable, the linear regression equation [13-3] from the text, has the form:

$$Y' = a + bX$$

For more than one independent variable, the equation is extended to include the additional variables. For k independent variables we use text formula [14-1]:

Multiple Regression Equation $$Y' = a + b_1X_1 + b_2X_2 + b_3X_3 + ... + b_kX_K \qquad [14-1]$$

Where:
X_1 is one of the independent variables.
X_2 is the second independent variable.
X_3 is the third independent variable.
X_k is the kth independent variable.
a is the Y-intercept, the value of Y when all the X's are zero.
b_j is the net change in Y' for each unit change in X_j, holding all other $X's$ constant.
j the subscript can assume values between 1 and k, which is the number of independent variables.

The values of b_1, and b_2, etc. are called the ***partial regression coefficients, net regression coefficients*** or just ***regression coefficients***. They indicate the change in the estimated value of the dependent variable for a unit change in one of the independent variables, when the other independent variables are held constant.

This equation can be extended for any number of independent variables.

For example, suppose the National Sales Manager of General Motors wants to analyze regional sales using the number of autos registered in the region (X_1), the average age of the automobiles registered in the region (X_2), and the personal income in the region (X_3). Some of the sample information obtained is:

Region	Sales ($ millions)	Number of autos in region (000)	Average age of autos (years)	Personal Income in Region (billions)
	Y	X_1	X_2	X_3
I	$9.2	842	5.6	$29.5
II	46.8	2,051	5.1	182.6
III	26.2	1,010	5.8	190.7

Suppose the multiple regression equation was computed to be:

$$Y' = 41.0 + 0.0071X_1 + (-3.19)X_2 + 0.01611X_3$$

In April of this year the automobile registration bureau announced that in a particular region 1,542,000 autos were registered, and their average age was 6.0 years. Another agency announced that personal income in the region was $150 billion. The sales manager could then estimate, as early as April, annual sales for this year by inserting the value of these independent variables in the equation and solving for Y'

$$\begin{aligned} Y' &= 41.0 + 0.0071(1{,}542) - 3.19(6.0) + 0.01611(150) \\ &= \$35.2 \text{ million} \end{aligned}$$

What is the meaning of the regression coefficients? The 0.0071 associated with number of autos in the region (in thousands) indicates that for each additional 1,000 autos registered, sales will increase 0.0071 (million), if the other independent variables are held constant. That is, the regression coefficients show change in the dependent variable when the other independent variables are not allowed to change.

Multiple Standard Error of Estimate

It is likely that there is some error in the estimation. This can be measured by the ***multiple standard error of estimate***.

Multiple standard error of estimate: Measures the error in the predicted value of the dependent variable.

Like the standard error of estimate described in the previous chapter, it is based on the squared deviations between Y and Y'. Text formula [14-2] is used.

Multiple Standard Error of Estimate

$$s_{y \cdot 12 \cdots k} = \sqrt{\frac{\Sigma(Y - Y')^2}{n - (k+1)}} \qquad [14-2]$$

Where:

Y is the observation.
Y' is the value estimated from the regression equation.
n is the number of observations in the sample.
k is the number of independent variables.
$s_{y \cdot 12 \cdots k}$ is the standard error of estimate. The subscripts indicate the number of independent variables being used to estimate the value of Y.

Assumptions About Multiple Regression and Correlation

As noted in previous chapters, it is generally considered good practice to identify the assumptions related to a topic because if the assumptions are not met fully, the results might be biased. There are five assumptions that must be met in multiple correlation.

1. There is a linear relationship between the dependent variable and each independent variable.
2. The dependent variable is continuous and at least interval scale.
3. The variation in the difference between the actual and the predicted values is the same for all fitted values of Y. That is, $(Y - Y')$ must be approximately the same for all values of Y'. When this is the case, differences exhibit ***homoscedasticity***.

   Homoscedasticity: The residuals are the same for all estimated values of the dependent variable.

4. The ***residuals***, computed by $(Y - Y')$, are normally distributed with a mean of 0.

   Residual: The difference between the actual and the predicted value of the dependent variable.

5. Successive observations of the dependent variable are uncorrelated. Violation of this assumption is called ***autocorrelation***. Autocorrelation often happens when data are collected successively over periods of time.

   Autocorrelation: Correlation of successive residuals. This condition frequently occurs when time is involved in the analysis.

Seldom in a real world example are all of the conditions met fully. However, the technique of regression still works effectively. If there is concern regarding the violation of one or more of the assumptions, it is suggested that a more advanced statistics book be consulted.

The ANOVA Table

A convenient means of showing the regression output is to use an ANOVA table. This table was first described in Chapter 12 and also mentioned in Chapter 13. The variation in the dependent variable is separated into two components: (1) that explained by the ***regression***, that is, the independent variable and (2) the ***residual error*** or unexplained variation.

These two categories are identified in the source column of the following ANOVA table. The column headed "*df*" refers to the degrees of freedom associated with each category. The total degrees of freedom is $(n - 1)$.

The degrees of freedom for regression is k, the number of independent variables. The degrees of freedom associated with the error term is $n - (k + 1)$. The SS in the middle of the top row of the ANOVA table refers to the sum of squares, or the variation.

$$\text{Total variation} = SS\,total = \Sigma\left(Y - \bar{Y}\right)^2$$

$$\text{Residual error} = SSE = \Sigma\left(Y - Y'\right)^2$$

$$\text{Regression variation} = SSR = \Sigma\left(Y' - \bar{Y}\right)^2 = \left(SS\,Total - SSE\right)$$

The column headed MS refers to the mean square and is obtained by dividing the SS term by the df term. Thus, MSR, the mean square regression, is equal to SSR/k, and MSE equals SSE/ $[n - (k + 1)]$. The general format of the ANOVA table is:

Analysis of Variance

Source	*df*	SS	MS	*F*
Regression	k	SSR	MSR = SSR/k	MSR / MSE
Error	$n - (k + 1)$	SSE	MSE = SSE/[$n - (k + 1)$]	
Total	$n - 1$	SS total		

Another measure of the effectiveness of the regression equation is the ***coefficient of multiple determination***.

> ***Coefficient of multiple determination***: The percent of the total variation in the dependent variable that is explained by the variation in the independent variables.

The coefficient of multiple determination, written R^2 or R square, may range from 0 to 1.0. It is the percent of the variation explained by the regression. The ANOVA table is used to calculate the coefficient of multiple determination. It is the sum of squares due to the regression divided by the sum of squares total.

Coefficient of Multiple Determination

$$R^2 = \frac{SSR}{SS\,total} \qquad [14-3]$$

In the Automobile Sales example, if the coefficient of multiple determination were 0.81, it would indicate that the three independent variables, considered jointly, explain 81 percent of the variation in millions of sales dollars.

The multiple standard error of estimate may also be found directly from the ANOVA table.

$$s_{y \cdot 1,2 \cdots k} = \sqrt{\frac{SSE}{n-(k+1)}}$$

Correlation Matrix

A ***correlation matrix*** is useful in analyzing the relationship between the dependent variable and each of the independent variables.

> ***Correlation matrix***: A matrix showing the coefficients of correlation between all pairs of variables.

The correlation matrix is also used to check for ***multicollinearity***.

> ***Multicollinearity***: Correlation among the independent variables.

Multicollinearity can distort the standard error of estimate and may, therefore, lead to incorrect conclusions as to which independent variables are statistically significant.

Global Test: Testing the Multiple Regression Model

The overall ability of the independent variables $X_1, X_2, \ldots X_k$, to explain the behavior of the dependent variable Y can be tested. Two tests of hypotheses are considered in this chapter. The first one is called the ***global test***.

> ***Global test***: A test used to investigate whether all the independent variables have zero net regression coefficients

It tests the overall ability of the set of independent variables to explain differences in the dependent variable. The null hypothesis is that the net regression coefficients in the population are all zero. If accepted, it would imply that the set of coefficients is of no value in explaining differences in the dependent variable. The alternate hypothesis is that *at least* one of the coefficients is not zero. This test is written in symbolic form for three independent variables as:

$$H_0: \beta_1 = \beta_2 = \beta_3 = 0$$
$$H_1: \text{Not all the } \beta \text{ 's are } 0$$

Rejecting H_0 and accepting H_1 implies that one or more of the independent variables is useful in explaining differences in the dependent variable. However, a word of caution, it does not suggest how many or identify which regression coefficients are not zero.

The test statistic used is the F distribution, which was first described in Chapter 12, the ANOVA chapter.

Recall these characteristics of the F distribution:

1. It is positively skewed, with the critical value located in the right tail. The critical value is the point that separates the region where H_0 is not rejected from the region of rejection.

2. It is constructed by knowing the number of degrees of freedom in the numerator and the number of degrees of freedom in the denominator.

To employ the F distribution, two sets of degrees of freedom are required. The degrees of freedom for the numerator are equal to k, the number of independent variables. The degrees of freedom in the denominator is equal to $n - (k + 1)$ where n refers, as usual, to the total number of observations.

The value of F is found by text formula [14-4]:

Global Test

$$F = \frac{MSR}{MSE} = \frac{SSR/k}{SSE/[n-(k+1)]} \quad [14-4]$$

Where:

SSR	is the sum of the squares "explained by" the regression.
k	is the number of independent variables.
SSE	is the sum of squares error.
n	is the number of observations.

Evaluating Individual Regression Coefficients

The second test of hypothesis identifies which of the set of independent variables are significant predictors of the dependent variable. That is, it tests the independent variables individually rather than as a unit. This test is useful because unimportant variables can be eliminated.

The test statistic is the Student t distribution with $n - (k + 1)$ degrees of freedom. For example, suppose we want to test whether the second independent variable is zero or greater, versus the alternate that it was less than zero. The null and alternate hypotheses would be written as follows.

$$H_0\text{: } \beta_2 \geq 0$$
$$H_1\text{: } \beta_2 < 0$$

Rejection of the null hypothesis and acceptance of the alternate hypothesis would imply that variable number two is significant and that it has an inverse relationship with the dependent variable. Obviously its sign is negative.

Qualitative Independent Variables

The variables used in regression analysis must be ***quantitative variables.***

Quantitative variable: A numeric variable that is at least interval scale.

Recall that a quantitative variable is a variable that is numerical in nature, such as, the number of hours worked by employees, the number of traffic accidents in Swanton in a week, and the distance traveled to work.

However, frequently we want to use variables that are not numeric, such as gender, whether a home has a computer, or whether an answer is yes or no. These variables are called ***qualitative variables.***

Qualitative variable: A nonnumeric variable.

Qualitative variables are also called ***dummy variables*** or indicator variables.

***Dummy variable*:** A variable in which there are only two possible outcomes. For analysis one of the outcomes is coded a 1 and the other a 0.

For example, we are interested in estimating the selling price of a used automobile based on its age. Selling price is the dependent variable and age is one independent variable. Another variable is whether or not the car was manufactured in the United States. Note that a particular car can assume only two conditions: either it was built in the U. S. or it was not. To employ a qualitative variable a 0 or 1 coding scheme is used. In the study regarding estimated selling prices of used automobiles, those made in the U.S. are coded 1 and all others as 0.

Analysis of Residuals

We described the assumptions for regression and correlation analysis as:

1. There is a linear relationship between the dependent variable and the independent variables
2. The dependent variable is measured as an interval scale or ratio scale variable.
3. Successive observations of the dependent variable are not correlated.
4. The residuals, the difference between actual values and the estimated values, are normally distributed.
5. The variation in the residuals is the same for all fitted values of Y'. That is, the distribution of $(Y - Y')$ is the same for all values of Y'.

The last two assumptions can be verified by plotting the residuals. We want to confirm that the residuals follow a normal distribution and that the residuals have the same variation whether the Y' value is large or small.

Glossary

***Autocorrelation*:** Correlation of successive residuals. This condition frequently occurs when time is involved in the analysis.

***Coefficient of multiple determination*:** The percent of the total variation in the dependent variable that is explained by variation in the independent variables.

***Correlation matrix*:** A matrix showing the coefficients of correlation between all pairs of variables.

***Dummy variable*:** A variable in which there are only two possible outcomes. For analysis one of the outcomes is coded a 1 and the other a 0.

***Global test*:** A test used to investigate whether all of the independent variables have zero net regression coefficients.

***Homoscedasticity*:** The residuals are the same for all estimated values of the dependent variable.

***Multicollinearity*:** Correlation among the independent variables.

***Multiple Regression*:** A set of techniques used to analyze the relationship between two or more independent variables and a dependent variable.

Multiple standard error of estimate: Measures the error in the predicted value of the dependent variable.

Qualitative variable: A nonnumeric variable.

Quantitative variable: A numeric variable that is at least interval scale.

Residual: The difference between the actual and the predicted value of the dependent variable.

Chapter Examples

Example 1 – The Correlation Matrix

The Skaff Appliance Company currently has over 1,000 retail outlets throughout the United States and Canada. They sell name brand electronic products, such as TVs, stereos, VCRs, DVD players, cell phones, and microwave ovens. Skaff Appliance is considering opening several additional stores in other large metropolitan areas.

Paul Skaff, president, would like to study the relationship between the sales at existing locations and several factors regarding the existing store or its region. The factors are the population and the unemployment in the region, and the advertising expense of the store. Another variable considered is "mall." Mall refers to whether the existing store is located in an enclosed shopping mall or not. A "1" indicates a mall location; a "0" indicates the store is not located in a mall. A random sample of 30 stores is selected.

Use the MINITAB system output to:

a. Determine which independent variable has the strongest correlation with sales.

b. Comment on multicollinearity.

c. Conduct a test of hypothesis to determine if any of the regression coefficients are not equal to zero.

Sales (000)	Population (000,000)	Percent Unemployed	Advertising Expense (000)	Mall Location
5.17	7.50	5.1	59.0	0
5.78	8.71	6.3	62.5	0
4.84	10.00	4.7	61.0	0
6.00	7.45	5.4	61.0	1
6.00	8.67	5.4	61.0	1
6.12	11.00	7.2	12.5	0
6.40	13.18	5.8	35.8	0
7.10	13.81	5.8	59.9	0
8.50	14.43	6.2	57.2	1
7.50	10.00	5.5	35.8	0
9.30	13.21	6.8	27.9	0
8.80	17.10	6.2	24.1	1
9.96	15.12	6.3	27.7	1
9.83	18.70	0.5	24.0	0
10.12	20.20	5.5	57.2	1
10.70	15.00	5.8	44.3	0
10.45	17.60	7.1	49.2	0
11.32	19.80	7.5	23.0	0
11.87	14.40	8.2	62.7	1
11.91	20.35	7.8	55.8	0
12.60	18.90	6.2	50.0	0
12.60	21.60	7.1	47.6	1
14.24	25.25	0.4	43.5	0
14.41	27.50	4.2	55.9	0
13.73	21.00	0.7	51.2	1
13.73	19.70	6.4	76.6	1
13.80	24.15	0.5	63.0	1
14.92	17.65	8.5	68.1	0
15.28	22.30	7.1	74.4	1
14.41	24.00	0.8	70.1	0

Solution 1 – The Correlation Matrix

The first step is to determine the correlation matrix. It shows all possible simple coefficients of correlation. The MINITAB output is as follows:

	Sales	Popul	%-unemp	Adv
Popul	0.894			
%-unemp	–0.198	–0.368		
Adv	0.279	0.125	–0.030	
Mall	0.155	0.085	0.017	0.259

a. Sales is the dependent variable. Of particular interest is which independent variable has the strongest correlation with sales. In this case it is population (0.894). The negative sign between sales and %-unemp indicates that as the unemployment rate increases, sales decrease.

b. A second use of the correlation matrix is to check for multicollinearity. Multicollinearity can distort the standard error of estimate and lead to incorrect conclusions regarding which independent variables are significant. The strongest correlation among the independent variables is between %-unemp and popul (–0.368). A rule of thumb is that a correlation between –0.70 and 0.70 will not cause problems and can be ignored. At this point it does not appear there is a problem with multicollinearity.

c. We want to test the overall ability of the set of independent variables to explain the behavior of the dependent variable. Do the independent variables, population, percent unemployed, advertising expense, and mall explain a significant amount of the variation in sales? This question can be answered by conducting a global test of the regression coefficients. The null and alternate hypotheses are

$$H_0: \beta_1 = \beta_2 = \beta_3 = \beta_4 = 0$$

H_1: At least one of the β's is not zero.

The null hypothesis states that the regression coefficients are all zero. If they are all zero, this indicates they are of no value in explaining differences in the sales of the various stores. If the null hypothesis is rejected and the alternate accepted the conclusion is that at least one of the regression coefficients is not zero. Hence, we would conclude that at least one of the variables is significant in terms of explaining differences in sales.

The *F* distribution introduced in Chapter 11 is used as the test statistic. The *F* distribution is based on the degrees of freedom in the numerator and in the denominator. The degrees of freedom associated with the regression, which is the numerator, is equal to the number of independent variables. In this case there are four independent variables, so there are 4 degrees of freedom in the numerator. The degrees of freedom in the error row is $n - (k + 1) = 30 - (4 + 1) = 25$. There are 25 degrees of freedom in the denominator. The critical value of *F* is obtained from Appendix G. Find the column with 4 degrees of freedom and the row with 25 degrees of freedom in the table for the 0.05 significance level. The value is 2.76. The null hypothesis is rejected if the computed *F* is greater than 2.76.

The output from MINITAB is as follows.

```
Analysis of Variance
Source          DF      SS         MS        F          P
Regression       4  270.461     67.615    34.71      0.000
Error           25   48.705      1.948
Total           29  319.166
```

The computed value of F is 34.71 as shown above. It is also computed using formula [14-4].

$$F = \frac{\dfrac{SSR}{k}}{\dfrac{SSE}{[n-(k+1)]}} = \frac{\dfrac{270.461}{4}}{\dfrac{48.705}{[30-(4+1)]}} = \frac{67.615}{1.948} = 34.71$$

Since the computed value of 34.71 exceeds the critical value of 2.76, the null hypothesis is rejected and the alternate accepted. The conclusion is that at least one of the regression coefficients does not equal zero.

Self-Review 14.1

Check your answers against those in the ANSWER section.

Todd Heffren, President of Heffren Manufacturing Co., is studying the power usage at his Vanengo Plant. He believes that the amount of electrical power used is a function of the outside temperature during the day, and the number of units produced that day. A random sample of ten days is selected. The power usage in thousands of kilowatt hours of electricity and the production on that date is obtained. The National Weather Service is contacted to obtain the high temperature for the selected dates.
The MINITAB system was used to compute a correlation matrix and ANOVA table for the Heffren Manufacturing Co. data.

Power Used	Temperature (F)	Units Produced
12	83	120
11	79	110
13	85	128
9	75	101
14	87	105
10	81	108
12	84	110
11	77	107
14	85	112
11	84	119

```
           Usage       Temp
Temp       0.838
Output     0.361       0.506

Analysis of Variance
SOURCE      DF        SS          MS
Regression   2     17.069       8.534
Error        7      7.031       1.004
Total        9     24.100
```

a. Do you see any problems with multicollinearity?
b. Test the hypothesis that all the regression equations are zero. Use the 0.05 significance level.

Example 2 – Comparing Regression Coefficients

In Example 1 we found that at least one of the four independent variables had a regression coefficient different from zero.

a. Use the MINITAB system to aid in determining which of the regression coefficients is not equal to zero.

b. Would you consider deleting any of the independent variables?

Solution 2 – Comparing Regression Coefficients

The following output is from MINITAB.

The regression equation is: Sales = –1.67 + 0.552 Popul + 0.203 %-unemp + 0.0314 Adv + 0.220 Mall

Predictor	Coef	Stdev	t-ratio	p
Constant	-1.669	1.408	-1.18	0.247
Popul	0.55191	0.05063	10.90	0.000
%-unemp	0.2032	0.1171	1.74	0.095
Adv	0.03135	0.01606	1.95	0.062
Mall	0.2198	0.5400	0.41	0.687

S = 1.396 R-Sq = 84.7% R-Sq(adj) = 82.3%

a. The four independent variables explain 84.7% of the variation in sales. For those coefficients where the null hypothesis that the regression coefficients are equal to zero cannot be rejected, we will consider eliminating them from the regression equation. We are actually conducting four tests of hypotheses

For Population	For % Unemployed	For Advertising	For Mall
$H_0: \beta_1 = 0$	$H_0: \beta_2 = 0$	$H_0: \beta_3 = 0$	$H_0: \beta_4 = 0$
$H_1: \beta_1 \neq 0$	$H_1: \beta_2 \neq 0$	$H_1: \beta_3 \neq 0$	$H_1: \beta_4 \neq 0$

We will use the 0.05 significance level and a two-tailed test. The test statistic is the t distribution with $n - (k + 1) = 30 - (4 + 1) = 25$ degrees of freedom. The decision rule is to reject the null hypothesis if the computed value of t is less than –2.060 or greater than 2.060.

From the MINITAB output, the column labeled "Coef" reports the regression coefficients. The "Stdev" column reports the standard deviation of the slope coefficients. The "t-ratio" column reports the computed value of the test statistic.

The t-ratio for population (10.90) exceeds the critical value, but the computed values for percent unemployed, (1.74) advertising expense (1.95), and mall (0.41) are not in the rejection region. This indicates that the independent variable population should be retained and the other three dropped.

b. However, there is a problem that occurs in many real situations. Note that both percent unemployed and advertising expense are close to being significant. In fact, advertising expense would be significant if we increased the level of significances to 0.10. (The critical value would be –1.708 and 1.708, and 1.95 is outside the critical region.)

Another indicator of trouble is a reversal of a sign of the regression coefficient. Earlier in Problem 1, in the correlation matrix, the correlation between percent unemployed and sales was negative. Note, in the above regression equation the sign of the coefficient is positive. (The regression coefficient is 0.203). A reversal of a sign such as this is often an indication of multicollinearity. The earlier conclusion that there was not a problem in this area should be reviewed. Perhaps one or both of the independent variables—either advertising expense or percent unemployment—should be included in the regression equation.

Self-Review 14.2

Check your answers against those in the ANSWER section.

Refer to Self-Review 14.1 The output is for the Heffren Manufacturing problem. The regression equation is shown below.

```
Usage = -16.8 + 0.37 Temp -0.0171  Output

Predictor    Coef        Stdev      t-ratio
Constant     -16.801     7.162      -2.35
Temp 7       0.037089    0.09962     3.72
Output       -0.01707    0.04791    -0.36
```

Conduct a test of hypothesis to determine which of the independent variables have regression coefficients not equal to zero. Use the 0.05 significance level.

Example 3 – Using a Revised Prediction Equation

The multiple regression and correlation data for Example 2 were rerun using the two most significant variables—population and advertising expense.

```
The regression equation is
      Sales = - 0.08 + 0.521 Popul + 0.0335 Adv

Predictor        Coef        StDev          T       P
Constant       -0.079        1.086      -0.07   0.943
Popul         0.52075      0.04806      10.83   0.000
Adv           0.03347     0.001590       2.10   0.045

S = 1.428        R-Sq = 82.8%       R-Sq(adj) = 81.5%

Analysis of Variance

Source        DF          SS          MS          F        P
Regression     2      264.12      132.06      64.78    0.000
Error         27       55.05        2.04
Total         29      319.17
```

a. What is the new multiple regression equation?

b. What is the Y' value for the first store?

c. What is the coefficient of multiple determination? Interpret.

d. Do the residuals approximate a normal distribution?

e. Are the residuals constant for all fitted values of Y' ?

Solution 3 – Using a Revised Prediction Equation

a. The regression equation is: Sales = – 0.08 + 0.521 Popul + 0.0335 Adv

$$Y' = -0.08 + 0.521X_1 + 0.0335X_3$$

b. Y' for the first store is found by substituting the value: $X_1 = 7.5$ and $X_3 = 59.0$ into the Y' equation.

$$Y' = -0.08 + 0.521X_1 + 0.0335X_3$$
$$= -0.08 + 0.521(7.5) + 0.0335(59) = 5.804$$

c. The coefficient of multiple determination is R-Sq on the printout. It is 82.8%. A total of 82.8% of the variation in sales is explained by the population and the advertising expense.

d. The MINITAB System was used to develop the fitted values of Y' and the residuals. The fitted values are obtained by substituting the actual values of population and advertising expense in the regression equation. For example, the first store was in a city having 7.5 million populations and advertising expense of 59.0 thousand dollars. We substituted these values in the regression equation and the estimated, or "fitted" values of Y' obtained was 5.804.

The MINITAB system will perform these time-consuming calculations. However, the results are slightly different, due to rounding. For example, MINITAB estimates 5.8017 for the first store compared to our estimate of 5.804.

The residual is the difference between the actual and the predicted value. For the first store it is –0.634, found by (Y – Y') = (5.17 – 5.804). The residuals are computed for the other 29 stores in a similar fashion.

Row	Sales Y	Fitted Y'	Residual $(Y - Y')$
1	5.17	5.8017	-0.63169
2	5.78	6.5489	-0.76894
3	4.84	7.1705	-2.33050
4	6.00	5.8426	0.15741
5	6.00	6.4779	–0.47790
6	6.12	6.0679	0.05213
7	6.40	7.9830	-1.58299
8	7.10	9.1177	-2.01773
9	8.50	9.3502	-0.85022
10	7.50	6.3270	1.17299
11	9.30	7.7342	1.56581
12	8.80	9.6327	-0.83270
13	9.96	8.7221	1.23788
14	9.83	10.4625	-0.63255
15	10.12	12.3549	-2.23493
16	10.70	9.2153	1.48474
17	10.45	10.7332	-0.28321
18	11.32	11.0019	0.31810
19	11.87	9.5187	2.35131
20	11.91	12.3862	-0.47618
21	12.60	11.4370	1.16304
22	12.60	12.7626	-0.16265
23	14.24	14.5261	-0.28614
24	14.41	16.1129	-1.70287
25	13.73	12.5707	1.15930
26	13.73	12.7439	0.98610
27	13.80	14.6060	-0.80601
28	14.92	11.3919	3.52814
29	15.28	14.0242	1.25579
30	14.41	14.7655	-0.35555

The residuals should approximate a normal distribution. The residuals from the right hand column above are organized into the following histogram. The shape seems to approximate the normal distribution.

```
Histogram of RESI1   N = 30
Midpoint    Count
    -2.5        1  *
    -2.0        2  **
    -1.5        2  **
    -1.0        4  ****
    -0.5        7  *******
     0.0        3  ***
     0.5        1  *
     1.0        5  *****
     1.5        3  ***
     2.0        0
     2.5        1  *
     3.0        0
     3.5        1
```

e. Another assumption, called homoscedasticity, requires that the residuals remain constant for all fitted values of Y'. A scatter diagram can be used to investigate. The horizontal axis is the fitted values, i.e., Y', and the vertical axis reflects the residuals. This assumption seems to be met, according to the following plot.

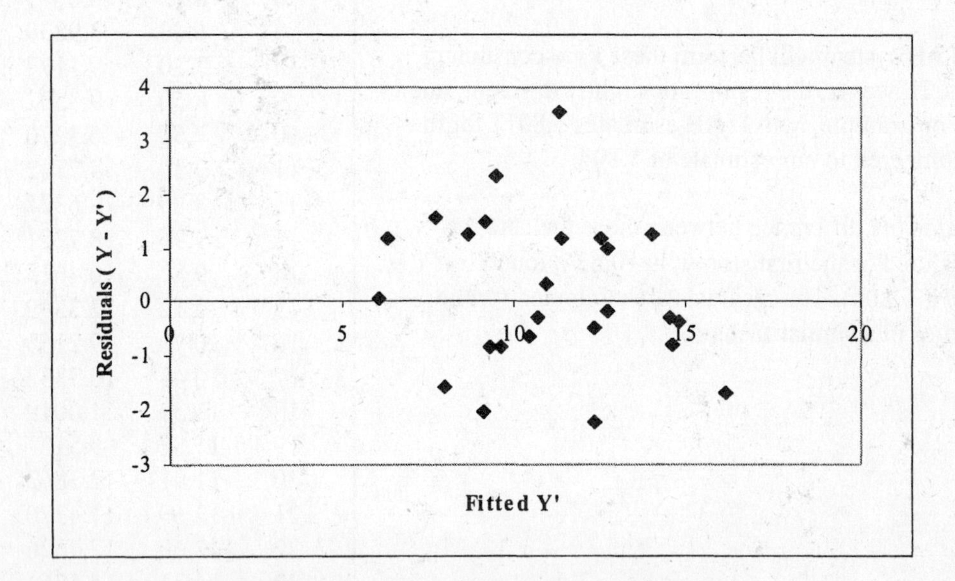

In summary, if Mr. Skaff wants to open new stores, the two variables that are the most effective in explaining differences in sales are the population in the surrounding area and the amount spent on advertising. The basic requirements for the use of regression analysis are met when these variables are used. The unemployment rate and whether or not the store is located in a mall are not important.

Self-Review 14.3

Check your answers against those in the ANSWER section.

The multiple regression and correlation data for Self-Reviews 14.1 and 14.2 were rerun using the most significant variable temperature. Use the MINITAB output shown below and on the next page to answer the following questions:

a. What is the new multiple regression equation?

b. What is the Y' value for the first item in the sample?

c. What is the coefficient of multiple determination? Interpret.

d. Do the residuals approximate a normal distribution?

e. Are the residuals constant for all fitted values of Y' ?

Using the independent variable temperature, the following regression equation was developed.

```
The regression equation is
Usage = - 17.2 + 0.353 Temp

Predictor        Coef        StDev          T        P
Constant      -17.241        6.658      -2.59    0.032
Temp          0.35294      0.08112       4.35    0.002

S = 0.9460        R-Sq = 70.3%      R-Sq(adj) = 66.6%

Analysis of Variance
Source        DF         SS         MS        F        P
Regression     1     16.941     16.941    18.93    0.002
Error          8      7.159      0.895
Total          9     24.100
```

continued on next page

Self-Review 14.3 Continued

The following two plots were obtained. Comment on the normality assumption and the condition of homoscedasticity.

```
Histogram of RESI2   N = 10
Midpoint   Count
    -1.6       1   *
    -1.2       1   *
    -0.8       0
    -0.4       2   **
     0.0       1   *
     0.4       3   ***
     0.8       0
     1.2       2   **
```

```
RESI2     -
          -
          -                                                X
    1.0+             X
          -
          -                                                          X
          -                    X
          -                                                X
    0.0+                                            X
          -  X
          -                                            X
          -
          -
   -1.0+
          -
          -                               X                X
          -
            ------+---------+---------+---------+---------+---------
+FITS2
               9.60     10.40     11.20     12.00     12.80     13.60
```

CHAPTER 14 ASSIGNMENT

MULTIPLE REGRESSION AND CORRELATION ANALYSIS

Name ______________________________ Section __________ Score______

Part I Select the correct answer and write the appropriate letter in the space provided.

______1. In a multiple regression equation there is more than one
- a. independent variable.
- b. dependent variable.
- c. coefficient of correlation.
- d. R^2 value.

______2. If the coefficient of multiple determination is 1, then the
- a. net regression coefficients are 0.
- b. standard error of estimate is 0.
- c. X values are also equal to 0.
- d. standard error of estimate is also 1.

______3. A dummy variable
- a. is also called an indicator variable.
- b. can only assume one of two values.
- c. is used as an independent variable.
- d. all of the above.

______4. In the global test of hypothesis
- a. we use the t distribution as the test statistic.
- b. we test to see if all of the net regression coefficients are 0.
- c. we test to insure that each of the independent variables is 0.
- d. all of the above.

______5. A residual is
- a. the independent variable.
- b. the dependent variable.
- c. the difference between the actual value and the fitted value.
- d. equal to R^2.

______6. The test for individual variables determines which independent variables
- a. have the most value in determining R^2.
- b. have nonzero regression coefficients.
- c. are used to compute the coefficients of correlation.
- d. are reported in the final equation.

______7. A correlation matrix shows the
- a. coefficients of correlation among all the variables.
- b. net regression coefficients.
- c. stepwise regression coefficients.
- d. residuals.

______8. Homoscedasticity refers to
- a. residuals that are correlated.
- b. independent variables that are correlated.
- c. a nonlinear relationship.
- d. residuals that are the same for all fitted values of Y'.

_____9. Multicollinearity means that
  a. the independent variables are correlated.
  b. time is involved with one of the independent variables.
  c. the dependent variable is correlated with the independent variables.
  d. the residuals do not have a constant variance.

_____10. When successive residuals are correlated we refer to this as
  a. multicollinearity. b. a dummy variable.
  c. autocorrelation. d. homoscedasticity.

Part II Record your answer in the space provided. Be sure to show essential calculations.

The information at the right is used for Problems 11 to 13:

William Clegg is the owner and CEO of Clegg QC Consulting. Mr. Clegg is concerned about the salary structure of his company and has asked the Human Relations Department to conduct a study. Mr. Stan Holt, an analyst in the department, is assigned the project. Stan selects a random sample of 15 employees and gathers information on the salary, number of years with Clegg Consulting, the employee's performance rating for the previous year, and the number of days absent last year.

Salary ($1000)	Years with Firm	Performance Rating	Days Absent
50.3	6	60	8
69.0	9	85	3
50.7	7	60	8
46.9	4	78	12
44.2	5	70	6
50.3	6	73	6
49.2	6	83	6
54.6	5	74	5
52.1	5	85	5
58.3	6	85	4
54.8	4	88	5
63.0	8	78	5
50.1	5	61	6
52.1	4	74	5
36.5	3	65	7

11. The correlation matrix at the right was developed from the MINITAB System.

	Salary	Years	Perform
Years	0.768		
Perform	0.514	0.130	
Absent	–0.587	–0.370	–0.435

  a. Which independent variable has the strongest correlation with salary?

a.

  b. Does the correlation matrix suggest any problems with the relationship among the variables?

____

12. Conduct a test of hypotheses to determine if any of the regression coefficients are not equal to 0. This analysis of variance table was computed as part of the output. Use the 0.05 significance level.

Analysis of Variance

SOURCE	*DF*	*SS*	*MS*	*F*	*P*
Regression	3	641.10	213.7	13.91	0.000
Error	11	168.94	15.36		
Total	14	810.04			

a. H_0: ______________________________

____

H_1: ______________________________

____

b. The decision rule is to reject H_0 if ______________________________

____

c. What is your decision?

Interpret it.

____

____

d. Determine the R-square value.

Interpret it.

____

____

13. Additional information was obtained from MINITAB. Conduct a test of hypothesis to determine if any of the regression coefficients do not equal 0. Use the 0.05 significance level. The regression equation is:

```
Salary = 19.2 + 3.10 Years + 0.269 Perform - 0.704 Absent
```

Predictor	Coef	Stdev	t-ratio	p
Constant	19.19	12.15	1.58	0.143
Years	3.0962	0.7061	4.38	0.001
Perform	0.2694	0.1196	2.25	0.046
Absent	-0.7043	0.5859	-1.20	0.255

a. H_0: ____________________

____

H_1: ____________________

____

H_0: ____________________

____

H_1: ____________________

____

H_0: ____________________

____

H_1: ____________________

____

b. The decision rules are to reject H_0 if ____________________

____

c. What is your decision? Interpret.

____

____

CHAPTER 15
CHI-SQUARE APPLICATIONS

Chapter Goals

After completing this chapter, you will be able to:

1. List the characteristics of the Chi-square distribution.
2. Conduct a test of hypothesis comparing an observed set of frequencies to an expected distribution.
3. Conduct a hypothesis test to determine whether two classification criteria are related.

Introduction

Recall that in Chapters 9 through 12, the data was interval or ratio scale such as weights of shipments of computers, income of McDonald's employees, or number of years on the job. The population from which the sample was drawn was assumed to be normal. What if these conditions cannot be met?

Recall that the nominal scale of measurement requires only that the sample information be categorized, with no order implied. As an example, students are classified by major, such as, business, history, computer science, etc.

This chapter considers a new test statistic, a chi-square distribution, where only the nominal level of measurement is required.

The Chi-Square Distribution

In the previous chapters the standard normal, the *t* distribution, and the *F* distributions were used as the test statistics. Recall that a test statistic is a quantity, determined from the sample information, used as a basis for deciding whether to reject the null hypothesis. In this chapter another distribution, called chi-square and designated χ^2, is used as the test statistic. It is similar to the *t* and *F* distributions in that there is a family of χ^2 distributions, each with a different shape, depending on the number of degrees of freedom. When the number of degrees of freedom is small the distribution is positively skewed, but as the number of degrees of freedom increases it becomes symmetrical and approaches the normal distribution. Chi-square is based on squared deviations between an observed frequency and an expected frequency and, therefore, it is always positive.

Goodness-of-Fit Tests

In the ***goodness-of-fit test*** the χ^2 distribution is used to determine how well an "observed" set of observations "fit" an "expected" set of observations.

> ***Goodness-of-fit test:*** A test designed to compare an observed distribution to an expected distribution.

The purpose of the goodness-of-fit test is to determine if there is a statistical difference between the two sets of data, one of which is observed and the other expected.

For example, an instructor told a class that the grading system would be "uniform." That is, that the same number of A's, B's, C's, D's and F's would be given. Suppose that the grades shown at the right were recorded at the end of the semester:

Grade	Number
A	12
B	24
C	23
D	30
F	11
	100

The question to be answered is: Do these final grades depart significantly from those that could be expected if the instructor had in fact graded uniformly? The null and alternate hypotheses are:

H_0: The distribution is uniform.
H_1: The distribution is not uniform.

The sampling distribution follows the χ^2 distribution and the value of the test statistic is computed by text formula [15-1]:

Chi-Square Test Statistic $$\chi^2 = \Sigma\left[\frac{(f_o - f_e)^2}{f_e}\right] \quad [15-1]$$

Where:
f_o is the observed frequency in a particular category.
f_e is the expected frequency in a particular category.
k is the number of categories with ($k - 1$) degrees of freedom.

It is not necessary that the expected frequencies be equal to apply the goodness-of-fit test. For example, at Scandia Technical Institute, over the years, 50 percent of the students were classified as freshmen, 40 percent sophomores, and 10 percent unclassified.

A sample of 200 students this past semester revealed that 90 were freshmen, 80 were sophomores, and 30 were unclassified. The null and alternate hypotheses are:

H_0: The distribution of students has not changed.
H_1: The distribution of students has changed.

Characteristics of the Chi-Square Distribution

The chi-square distribution has the following characteristics:

1. **Chi-square values are never negative.**
   This is because the difference between f_o and f_e is squared, that is, $(f_o - f_e)^2$.

2. **There is a family of chi-square distributions.**
   There is a chi-square distribution for one degree of freedom, another for two degrees of freedom, another for three degrees of freedom, and so on.

3. **The chi-square distribution is positively skewed.**
   However, as the number of degrees of freedom increases, the distribution begins to approximate the normal distribution.

Contingency Table Analysis

The χ^2 distribution is also used to determine if there is a relationship between two criteria of classification. As an example, we are interested in whether there is a relationship between job advancement within a company and the gender of the employee. A sample of 100 employees is selected. The survey results are shown in the table on the right.

Gender	No Advancement	Slow Advancement	Rapid Advancement	Total
Male	7	13	30	50
Female	13	17	20	50
Total	20	30	50	100

Note that an employee is classified two ways: by gender and by advancement. When an individual or item is classified according to two criteria, the resulting table is called a contingency table.

> ***Contingency table***: A two-way classification of a particular observation in table form.

The null and alternate hypotheses are expressed as follows:

H_0: There is no relationship between gender and advancement.
H_1: There is a relationship between gender and advancement.

Formula [15-1] is used to compute the value of the test statistic. The expected frequency, f_e is computed by noting that 50/100 or 50 percent of the sample is male. If the null hypothesis is true and advancement is unrelated to the gender of the employee, then it is expected that 50 percent of those who have not advanced will be male. The expected frequency, f_e, for males who have not advanced is 10, found by (0.50)(20). The other expected frequencies are computed similarly. If the difference between the observed and the expected value is too large to have occurred by chance, the null hypothesis is rejected.

Limitations of Chi-Square

There is a limitation to the use of the χ^2 distribution. If there is an unusually small expected frequency in a cell, chi-square (if applied) might result in an erroneous conclusion. This can happen because f_e appears in the denominator, and dividing by a very small number makes the quotient quite large! The value of f_e should be at least 5 for each cell (box). This requirement is to prevent any cell from carrying an inordinate amount of weight and causing the null hypothesis to be rejected.

Two generally accepted rules regarding small cell frequencies are:

1. If there are only two cells, the *expected frequency* in each cell should be 5 or more.

2. For more than two cells, chi-square should not be used if more than 20 percent of the f_e cells have expected frequencies less than 5.

Glossary

Contingency table: A two-way classification of a particular observation in table form.

Goodness-of-fit test: A test designed to compare an observed distribution to an expected distribution.

Chapter Examples

Example 1 – Computing the Value of Chi Square

A distributor of personal computers has five locations in the city of Ashland. The sales in units for the first quarter of the year are given in the table at the right. At the 0.01 significance level, do the records suggest that sales are uniformly distributed among the five locations?

Location	Sales (Units)
North Side	70
Pleasant Township	75
Southwyck	70
I-90	50
Venice Ave.	35
	300

Solution 1 – Computing the Value of Chi Square

The first step is to state the null hypothesis and the alternate hypothesis. The null hypothesis is that sales are uniformly distributed. The alternate is that there has been a change and the sales pattern is not uniformly distributed among the five stores. These hypotheses are written as follows:

H_0: Sales are uniformly distributed among the five locations
H_1: Sales are not uniformly distributed among the five locations

The appropriate test statistic is the χ^2 distribution. The critical value is obtained from Appendix B. The number of degrees of freedom is equal to the number of categories minus 1. There are five categories (locations), therefore there are four degrees of freedom found by $(k-1) = (5-1) = 4$. The problem states beforehand that the 0.01 significance level is to be used. To locate the critical value, find the column headed 0.01 and the row where *df*, the degree of freedom, is 4. The value at the intersection of this row and column is 13.277. Therefore, the decision rule is: Reject H_0 if the computed value of the test statistic exceeds 13.277, otherwise do not reject the null hypothesis. Graphically, the decision rule is shown on the right.

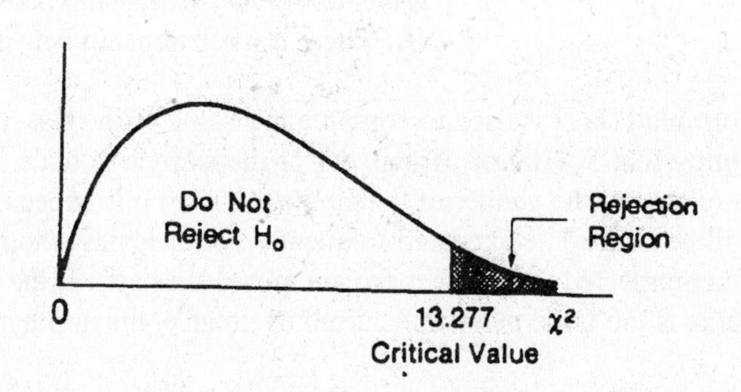

The observed frequencies, f_o are in Column 1 of the following table. The expected frequencies are in Column 2. How are the expected frequencies determined? If the null hypothesis is true (that sales are uniformly distributed among the five locations), then 1/5 of 300, or 60 computers should have been sold at each location.

	Col. 1	Col. 2	Col. 3	Col. 4	Col. 5
Location	f_o	f_e	$f_o - f_e$	$(f_o - f_e)^2$	$\frac{(f_o - f_e)^2}{f_e}$
North Side	70	60	10	100	1.67
Pleasant Township	75	60	15	225	3.75
Southwyck	70	60	10	100	1.67
I-90	50	60	−10	100	1.67
Venice Ave.	35	60	−25	625	10.42
	300	300	0		19.18

Recall that the value of the test statistic is computed by formula [15-1].

$$\chi^2 = \Sigma\left[\frac{(f_o - f_e)^2}{f_e}\right]$$

The value of the test statistic is determined by first taking the difference between the observed frequency and the expected frequency (Col. 3). Next these differences are squared (Col. 4). Then the result is divided by the expected frequency (Col. 5). This result is then summed over the five locations. The total is 19.18. The value of 19.18 is compared to the critical value of 13.277. Since 19.18 is greater than the critical value, H_0 is rejected and H_1 accepted. We conclude that sales are not uniformly distributed among the five locations.

Self-Review 15.1

Check your answers against those in the ANSWER section.

A tire manufacturer is studying the position of tires in blowouts. It seems logical that the tire blowouts will be uniformly distributed among the four positions. For a sample of 100 tire failures, is there any significant difference in that tire's position on the car? Use the 0.05 significance level.

Location of Tire on the Car

Left Front	Left Rear	Right Front	Right Rear
28	20	29	23

Example 2 – Computing the Value of Chi Square

From past experience the manager of the parking facilities at a major airport knows that 58 percent of the customers stay less than one hour, 23 percent between one and two hours, 10 percent between two and three hours, and nine percent three hours or more.

The manager wants to update this study. A sample of 500 stamped parking tickets is selected. The results showed 300 stayed less than one hour, 100 from one to two hours, 60 from two to three hours, and 40 parked three hours or more. At the 0.01 significance level does the data suggest there has been a change in the distribution of the length of time customers use the parking facilities?

Solution 2 – Computing the Value of Chi Square

The first step is to state the null hypothesis and alternate hypothesis.

H_0: There has been no change in the distribution of parking times.
H_1: There has been a change in the distribution of parking times

The next step is to determine the decision rule. (Note in the table below that there are four categories). The number of degrees of freedom is the number of categories minus 1. In this problem it is $(4 - 1) = 3$ degrees of freedom. Referring to Appendix I, the 0.01 level and 3 degrees of freedom, the critical value of chi-square is 11.345, so H_0 is rejected if χ^2 is greater than 11.345.

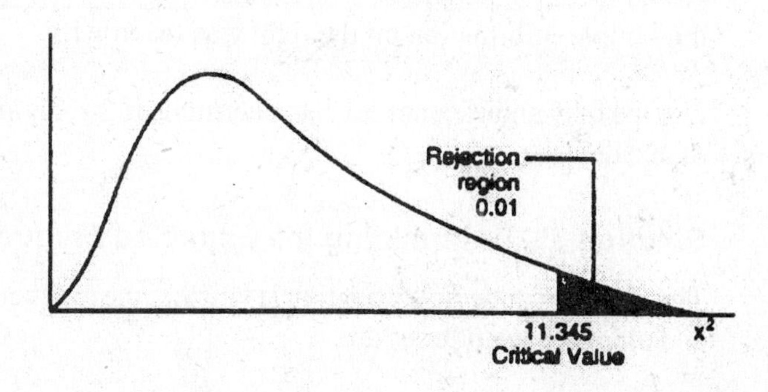

The value of the test statistic is computed as follows:

The observed frequencies from the sample are shown In Column 2 of the following table. Recall that based on past experience 58 percent of the customers parked their car less than one hour. If the null hypothesis is true, then 58% × 500 (in the sample) = 290, the expected frequency. Likewise, 23 percent stayed from one to two hours. Thus, 23% × 500 gives the expected frequency of 115. The complete set of expected frequencies is given in Column 3. Chi-square is computed to be 4.86.

	Col. 1	Col. 2	Col. 3	Col. 4	Col. 5	Col. 6
Time in Parking Lot	Percent of Total	Number in Sample f_o	f_e	$f_o - f_e$	$(f_o - f_e)^2$	$\frac{(f_o - f_e)^2}{f_e}$
Less than 1 hour	58	300	290	10	100	100/290 = 0.34
1 up to 2 hours	23	100	115	−15	225	225/115 = 1.96
2 up to 3 hours	10	60	50	10	100	100/50 = 2.00
3 hours or more	9	40	45	−5	25	25/45 = 0.56
Total	100	500	500	0		$\chi^2 = 4.86$

Since the computed value of chi-square (4.86) is less than the critical value (11.345), the null hypothesis is not rejected. There has been no change in the distribution of the lengths of parking time at the airport.

Self-Review 15.2

Check your answers against those in the ANSWER section.

In recent years, 42 percent of the American made automobiles sold in the United States were manufactured by General Motors, 33 percent by Ford, 22 percent by Damlier-Chrysler, and 3 percent by all others. A sample of the sales of American-made automobiles conducted last week revealed that 174 were manufactured by DamlierChrysler, 275 by Ford, 330 by GM, and 21 by all others.

Test the hypothesis at the 0.05 level that there has been no change in the sales pattern.

Example 3 – Determining the Expected Frequency and Chi Square

A study is made by an auto insurance company to determine if there is a relationship between the driver's age and the number of automobile accident claims submitted during a one-year period. From a sample of 300 claims, the sample information on the right was recorded.

	Age (Years)			
No. of Accidents	Less than 25	25-50	Over 50	Total
0	37	101	74	212
1	16	15	28	59
2 or more	7	9	13	29
Total	60	125	115	300

Use the 0.05 significance level to determine if there is any relationship between the driver's age and the number of accidents.

Solution 3 – Determining the Expected Frequency and Chi Square

The question under investigation is whether the number of auto accidents is related to the driver's age. The null and alternate hypotheses are.

H_0: There is no relationship between age and the number of accidents.
H_1: There is a relationship between age and the number of accidents.

The critical value is obtained from the chi-square distribution in Appendix H. The number of degrees of freedom is equal to the number of rows minus one, times the number of columns minus one. Hence, the degrees of freedom are (3 – 1) (3 – 1) = 4. The significance level, as stated in the problem, is 0.05. The critical value from Appendix I is 9.488. The null hypothesis is rejected if the computed value of χ^2 is greater than 9.488.

Formula [15-1], as cited earlier, is used to determine χ^2.

$$\chi^2 = \Sigma\left[\frac{(f_o - f_e)^2}{f_e}\right]$$

Where:
f_o is the frequency observed.
f_e the expected frequency.

The first step is to determine the expected frequency for each corresponding observed frequency. If the null hypothesis is true (the number of accidents is not related to age), we can expect 212 out of the 300 sampled, or 70.67 percent of the drivers to have had no accidents. Thus, we can expect 70.67 percent of the 60 drivers under 25 years, or 42.40 drivers, to have had no accidents.

Likewise, if the null hypothesis is true, 70.67 percent of the 125 drivers in the 25 to 50 age bracket, or 88.33 drivers, should have had no accidents.

The table at the right shows the complete set of observed and expected frequencies. Note that some totals were rounded.

	Age						
No. of Accidents	Less than 25		25-50		Over 50		Total
	f_o	f_e	f_o	f_e	f_o	f_e	
0	37	42.40	101	88.33	74	81.27	212
1	16	11.80	15	24.58	28	22.62	59
2 or more	7	5.80	9	12.08	13	11.12	29
Total	60	60.00	125	124.99	115	115.01	300

Note: The f_e values were calculated without rounding the percents. Ex: 212/300 = 70.666666%

The expected frequency for any category is found by text formula [15-2]:

$$\text{Expected frequency} = \frac{(\text{row total})(\text{column total})}{\text{grand total}}$$

The value for the first row and column is used as an example. There are 212 people who did not have any accidents, 60 persons are less than 25 years old, and there is a total of 300 people. These values are inserted into the formula:

$$f_e = \frac{(\text{row total})(\text{column total})}{\text{grand total}} = \frac{(212)(60)}{300} = 42.40$$

This is the same value computed previously.

The computed value of χ^2 is 11.03.

$$\chi^2 = \Sigma\left[\frac{(f_o - f_e)^2}{f_e}\right] = \frac{(37.00-42.40)^2}{42.40} + \frac{(101.00-88.33)^2}{88.33} + \ldots + \frac{(13-11.12)^2}{11.12} = 11.03$$

Since the computed value of χ^2 is greater than the critical value of 9.488, the null hypothesis is rejected and the alternate accepted. We conclude that there is a relationship between age and the number of accidents.

Self-Review 15.3

Check your answers against those in the ANSWER section.

A random sample of 480 male and female adults was asked the amount of time each person spent watching TV last week. Their responses are shown at the right. At the 0.05 significance level, does it appear that the amount of time spent watching TV is related to the gender of the viewer?

	Gender of Viewer		
Hours	**Male**	**Female**	**Total**
Under 8	70	90	160
8 to 15	100	60	160
15 or more	55	105	160
	225	255	480

CHAPTER 15 ASSIGNMENT

CHI-SQUARE APPLICATIONS

Name ____________________________ Section ________ Score ______

Part I Select the correct answer and write the appropriate letter in the space provided.

______1. In a goodness-of-fit test
- **a.** the sample size must be at least 30.
- **b.** no assumption is made regarding the shape of the population.
- **c.** σ is always known.
- **d.** the interval scale of measurement is required.

______2. The level of measurement required for the goodness-of-fit test is
- **a.** nominal.
- **b.** ordinal.
- **c.** interval.
- **d.** ratio.

______3. The chi-square distribution is
- **a.** positively skewed.
- **b.** a continuous distribution.
- **c.** based on the number of categories.
- **d.** all of the above.

______4. A two-way classification of the data is called a
- **a.** chi-square distribution.
- **b.** normal distribution.
- **c.** contingency table.
- **d.** none of the above.

______5. In a goodness-of-fit test where f_e values are the same in all four categories
- **a.** the degrees of freedom is 3.
- **b.** $k = 4$.
- **c.** the null hypothesis is that the proportion in each category is the same.
- **d.** all of the above.

______6. A sample of 100 undergraduate students is classified by major (3 groups) and gender. How many degrees of freedom are there in the test?
- **a.** 2
- **b.** 3
- **c.** 4
- **d.** 99

______7. In a chi-square test the $df = 4$. At the 0.05 significance level the critical value of the chi-square is
- **a.** 7.779
- **b.** 7.815
- **c.** 9.488
- **d.** 13.388

______8. The shape of the chi-square distribution is
- **a.** based on the degrees of freedom.
- **b.** based on the level of measurement.
- **c.** based on the shape of the population.
- **d.** based on at least 30 observations.

______9. In a test to find out if the two criteria of classification are related, the expected cell frequencies should be
- **a.** all less than 5.
- **b.** at least one less than 5.
- **c.** at least 5 percent of the population.
- **d.** at least 5.

_____ **10.** The sum of the observed and the expected frequencies are

- **a.** always at least 30.
- **b.** always the same.
- **c.** always less than 5 percent.
- **d.** always less than 5.

Part II Record your answer in the space provided. Show essential work.

11. The manager of a Farmer Jack Super Market would like to know if there is a preference for the day of the week on which customers do their shopping. A sample of 420 families revealed the following. At the 0.05 significance level, is there a difference in the proportion of customers that prefer each day of the week?

Day of the week	Number of persons
Monday	20
Tuesday	30
Wednesday	20
Thursday	60
Friday	80
Saturday	130
Sunday	80

a. State the null and alternate hypotheses.

H_0: ____________________

H_1: ____________________

b. State the decision rule.

____

c. Compute the value of the test statistic.

c.

Day of week				
Monday				
Tuesday				
Wednesday				
Thursday				
Friday				
Saturday				
Sunday				
Total				

d. What is your decision regarding the null hypothesis? Interpret the result.

____

____

12. A charity solicits donations by phone. From long experience the charity's director reports that 60 percent of the calls will result in refusal to donate, 30 percent will request more information via the mail, and 10 percent will result in an immediate credit card donation. For a sample of 200 calls last week, 140 refused to donate, 50 requested additional information, and 10 made an immediate donation. At the 0.10 significance level was the sample result different from the usual pattern?

a. State the null and alternate hypotheses.

H_0: ____

H_1: ____

b. State the decision rule.

____

____

c. Compute the value of the test statistic.

c.

d. What is your decision regarding the null hypothesis? Interpret the result.

____

____

____

13. There are three loan officers at Farmer National Bank. All decisions on mortgage loans are made by one of these officers. The president of the bank would like to be sure that the rejection rate is about the same for the three officers. A sample of 200 recent applications yielded the following results. Is the rejection rate related to the officer that processes the loan? Use the 0.05 significance level.

	Loan Officer		
	Felix	Otis	Foxburrow
Approved	50	70	55
Rejected	10	10	5
Total	60	80	60

a. State the null and alternate hypotheses.

H_0: ______________________________

H_1: ______________________________

b. State the decision rule.

____

____

c. Compute the value of the test statistic.

c.

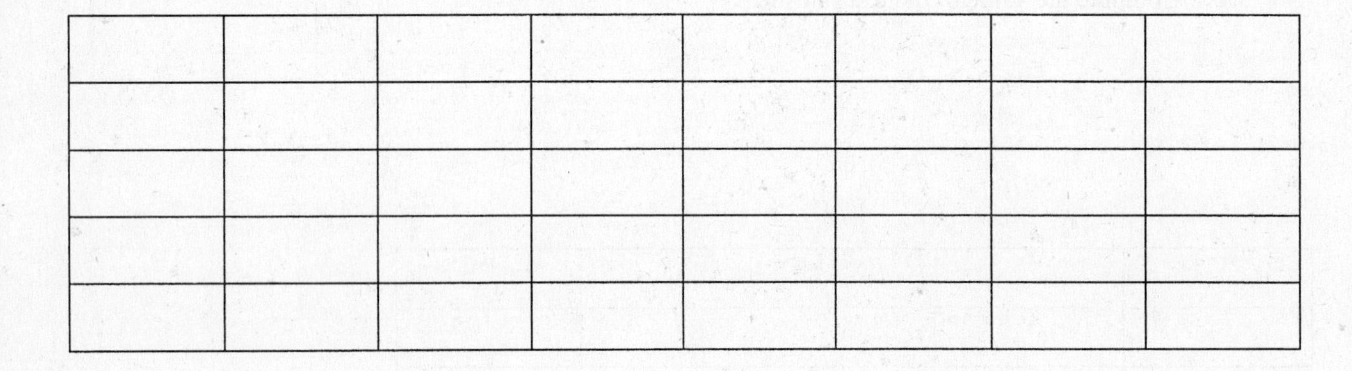

d. What is your decision regarding the null hypothesis? Interpret the result.

____

____

SELF-REVIEW ANSWER SECTION

CHAPTER 2

DESCRIBING DATA: FREQUENCY DISTRIBUTIONS AND GRAPHIC PRESENTATION

Self-Review 2.1

a. Number of classes: $2^k \geq n$, $2^5 \geq 20$, thus the number of classes is 5.

b. Class interval is 5: $i \geq \dfrac{H-L}{k}, \dfrac{39-18}{5} = \dfrac{21}{5} = 4.2$, thus i is > 4.2 (round up to 5)

c. Frequency distribution:

Miles per Gallon	Tallies	Number of Engines
15 up to 20	//	2
20 up to 25	////	4
25 up to 30	~~////~~ /	6
30 up to 35	~~////~~	5
35 up to 40	///	3
Total		20

Self-Review 2.2

Relative frequency distribution

Miles per Gallon	Tallies	Number of Engines	Relative Frequency	Found By
15 up to 20	//	2	10%	2/20
20 up to 25	////	4	20%	4/20
25 up to 30	~~////~~ /	6	30%	6/20
30 up to 35	~~////~~	5	25%	5/20
35 up to 40	///	3	15%	3/20
Total		20	100%	

Self-Review 2.3

Histogram

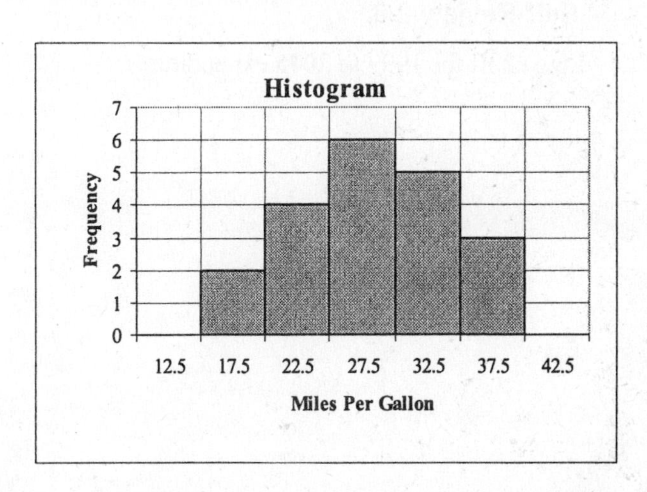

Self-Review 2.4

Frequency polygon

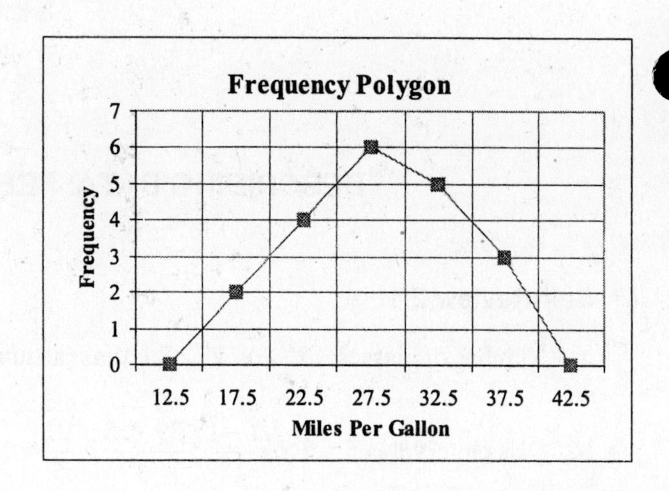

Self-Review 2.5

a. Cumulative Frequency Polygon

Number	Cumulative Total
2	2
4	6
6	12
5	17
3	20

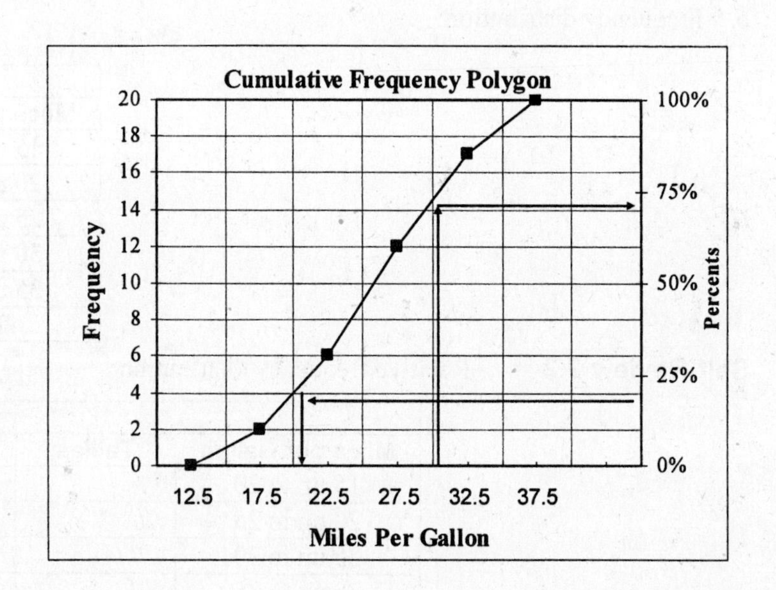

b. About 70% of the automobiles are getting less than 30 miles per gallon

c. About 20% of the automobiles are getting 20 miles per gallon or less

Self-Review 2.6.

Line chart for 1999 to 2005 expenditures.

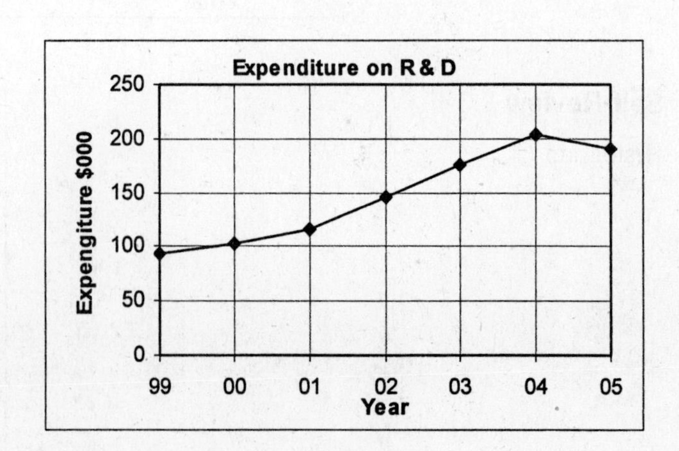

Self-Review 2.7.

Bar chart for 1999 to 2005 expenditures

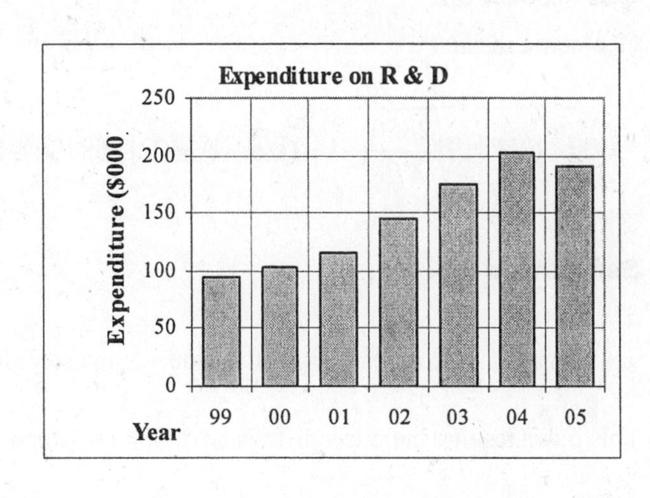

Self-Review 2.8.

Pie chart for new cars sold

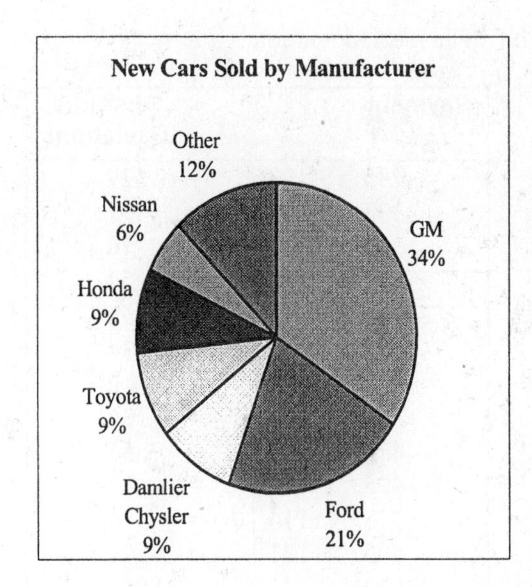

CHAPTER 3

DESCRIBING DATA: NUMERICAL MEASURES

Self-Review 3.1

a. Mean: $\overline{X} = \frac{\Sigma X}{n} = \frac{8+5+4+10+8+3+4}{7} = \frac{42}{7} = 6$

b. Median = 5. Middle data value. 3, 4, 4, 5, 8, 8, 10 (⇑ under 5)

c. Mode = 4 and 8

Self-Review 3.2

Geometric mean:

$$GM = \sqrt[15]{\frac{2850}{475}} - 1 = (\sqrt[15]{6} - 1) = (1.1268776 - 1) = 0.1268776 = 12.7\%$$

Self-Review 3.3

a. Range: $\text{Range} = \text{Highest Value} - \text{Lowest Value} = (\$595 - \$295) = \$300$

This indicates that there is a difference of $300 between the largest and the smallest monthly rent.

b. The mean deviation:

Payment X	$\lvert X - \overline{X} \rvert$		Absolute Deviations
295	|−129|	=	129
333	|−91|	=	91
345	|−79|	=	79
370	|−54|	=	54
370	|−54|	=	54
390	|−34|	=	34
422	|−2|	=	2
460	|+36|	=	36
475	|+51|	=	51
495	|+71|	=	71
538	|+114|	=	114
595	|+171|	=	171
5,088			$886

$$\overline{X} = \frac{\Sigma X}{n} = \frac{\$5,088}{12} = \$424$$

$$MD = \frac{\Sigma \lvert X - \overline{X} \rvert}{n} = \frac{\$886}{12} = \$74$$

The mean deviation of $74 indicates that the typical monthly rent deviates $74 from the mean of $424.

Self-Review 3.4

Weighted mean:

$$\overline{X} = \frac{w_1 X_1 + w_2 X_2 + w_3 X_3 + w_4 X_4}{w_1 + w_2 + w_{31} + w_4}$$

$$= \frac{31(\$30) + 42(\$10) + 47(\$20) + 63\$24)}{31 + 42 + 47 + 63} = \frac{\$3802}{183} = \$20.78$$

Self-Review 3.5

a. Range = \$27 - \$10 = \$17

b. Mean $= \overline{X} = \frac{\Sigma X}{n} = \frac{80}{5} = \$16$

c. Variance

$$s^2 = \frac{\Sigma(X-\overline{X})^2}{n-1} = \frac{174}{5-1} = \frac{174}{4} = 43.5$$

d. Standard deviation $s = \sqrt{43.5} = 6.595 = \$6.60$

X	$(X-\overline{X})$			$(X-\overline{X})^2$
10	10 – 16	=	–6	36
12	12 – 16	=	–4	16
15	15 – 16	=	–1	1
16	16 – 16	=	0	0
27	27 – 16	=	11	121
80				174

Self-Review 3.6

a. Percent:

$$k = \frac{X-\overline{X}}{s} = \frac{\$885-\$990}{\$70} = -1.5$$

$$k = \frac{X-\overline{X}}{s} = \frac{\$1095-\$990}{\$70} = 1.5$$

Applying Chebyshev's Theorem: $1-\frac{1}{k^2} = (1-\frac{1}{1.5^2}) = (1-\frac{1}{2.25}) = (1-0.4444) = 0.5556 = 55.6\%$

This means that at least 56% of the salespersons earn between \$885 and \$1095 in commission.

b. Positive skewness since the mean 990 is larger than the median 950.

CHAPTER 4

DESCRIBING DATA: DISPLAYING AND EXPLORING DATA

Self-Review 4.1

a. Sort

18	20	21	22	22	24	25	25	25	26	26	28	28	28	29	30	35	35	35	37

b. Dot plot.

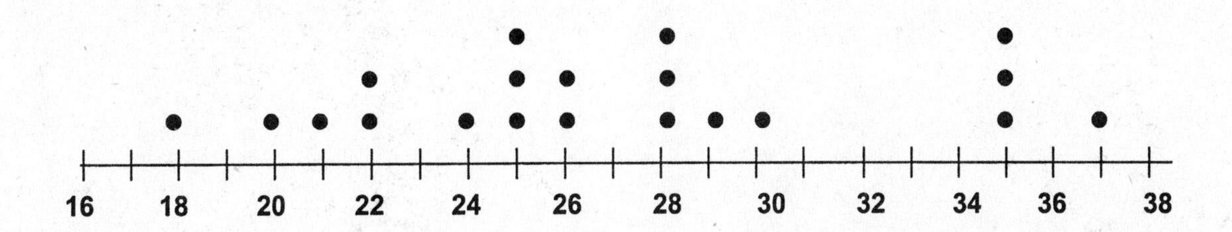

Self-Review 4.2

The data is first put into an ordered array.

25	28	39	50	61	65	81	82	85	85	85	86	90	92	120	137	140	142	148

a. First Quartile: Let $P = 25$ and

$$L_p = (n+1)\frac{P}{100} = (19+1)\frac{25}{100} = 5$$

Then locate the 5th observation in the array, which is 61. Thus $Q_1 = 61$ or $61,000.

b. Third quartile: Let $P = 75$ and

$$L_p = (n+1)\frac{P}{100} = (19+1)\frac{75}{100} = 15$$

Then locate the 15th observation in the array, which is 120. Thus $Q_3 = 120$ or $120,000.

c. The median: Let $P = 50$ and

$$L_p = (n+1)\frac{P}{100} = (19+1)\frac{50}{100} = 10$$

Then locate the 10th observation in the array, which is 85. Thus Q_2 = the median = 85 or $85,000.

d. Box plot: The five essential pieces of data are:

Minimum value = 25, $Q_1 = 61$, $Q_2 = 85$ $Q_3 = 120$, Maximum value = 148

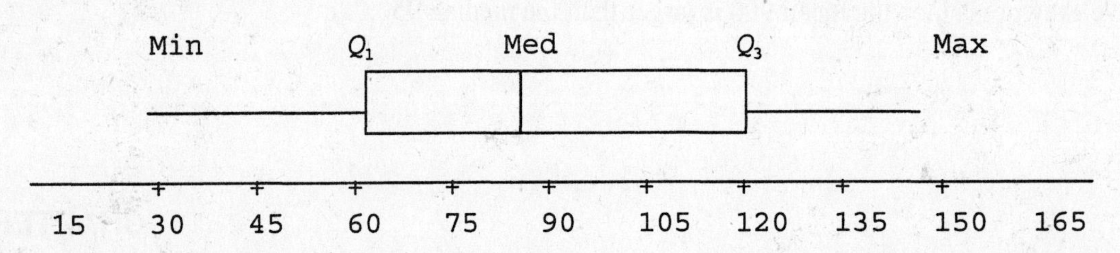

Self-Review 4.3

a. Positive skewness since the mean 1385 is larger than the median 1330.

b. Coefficient of skewness: $sk = \dfrac{3(\overline{X} - \text{median})}{s} = \dfrac{3(1385 - 1330)}{75} = 2.2$

Self-Review 4.4

a. Scatter diagram:

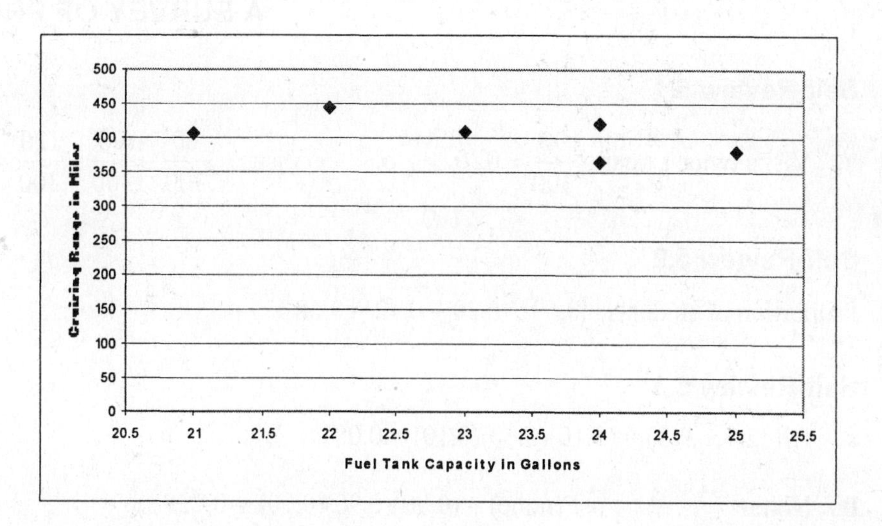

b. There is a slight decline in cruising range as tank capacity increases.

Self-Review 4.5

a. Percent males working at home = $\frac{8}{20} = 0.40 = 40\%$

b. Percent males working at office = $\frac{12}{20} = 0.60 = 60\%$

c. Percent employees working at office = $\frac{18}{40} = 0.45 = 45\%$

CHAPTER 5

A SURVEY OF PROBABILITY CONCEPTS

Self-Review 5.1

a. Visits twice a year: $\frac{90}{300} = 0.30$ **b.** Visits: $\frac{60}{300} + \frac{90}{300} + \frac{120}{300} = \frac{270}{300} = 0.90$

Self-Review 5.2

Proportion of students: $0.60 + 0.20 - 0.12 = 0.68$

Self-Review 5.3

a. All three: $P(3) = (0.10)(0.10)(0.10) = 0.001$

b. None: $P(\text{none}) = (0.90)(0.90)(0.90) = 0.729$

c. At least one: $1 - P(\text{none}) = (1 - 0.729) = 0.271$

Self-Review 5.4

a. Both female: $P(\text{both female}) = \frac{6}{10} \times \frac{5}{9} = 0.33$

b. At least one male: $P(\text{At least one male}) = \left(\frac{4}{10} \times \frac{3}{9}\right) + \left(\frac{4}{10} \times \frac{6}{9}\right) + \left(\frac{6}{10} \times \frac{4}{9}\right) = 0.67$

$m \; m \qquad m \; f \qquad f \; m$

Self-Review 5.5

a. Had Heart attack or is heavy smoker:

$$P(\text{Heart attack or heavy smoker}) = \frac{180}{500} + \frac{125}{500} - \frac{90}{500} = \frac{215}{500} = 0.43$$

b. Heavy Smoker and no heart attack: $P(\text{Heavy smoker and no heart attack}) = \frac{35}{500} = 0.07$

Self-Review 5.6

Number of different flights: $5 \times 10 = 50$

Self-Review 5.7

Number of different ways: $_nP_r = \frac{n!}{(n-r)!} = \frac{9!}{(9-3)!} = \frac{9!}{6!} = \frac{9 \times 8 \times 7 \times 6!}{6!} = 9 \times 8 \times 7 = 504$

Self-Review 5.8

Number of different trip combinations: $_nC_r = \frac{n!}{r!(n-r)!} = \frac{8!}{5!3!} = \frac{8 \times 7 \times 6 \times 5!}{5! \times 3 \times 2} = \frac{8 \times 7 \times 6}{3 \times 2} = 56$

DISCRETE PROBABILITY DISTRIBUTIONS

Self-Review 6.1

Number of Accidents per Month X	Probability $P(X)$	$xP(x)$	$(X-\mu)$	$(X-\mu)^2 P(X)$
0	0.60	0	$0-0.5$	$(0.25)(0.6)=0.150$
1	0.30	0.3	$1-0.5$	$(0.25)(0.3)=0.075$
2	0.10	0.2	$2-0.5$	$(2.25)(0.1)=0.225$
Σ	1.00	0.5		$\Sigma=0.450$

a. Mean: $\mu=\Sigma[xP(x)]=0.5$ b. Variance: $\sigma^2=\Sigma[(x-\mu)^2P(x)]=0.45$

Self-Review 6.2

a. Rules of probability: *R,R,NR* (0.60)(0.60)(0.40) = 0.144

R,NR,R (0.60)(0.40)(0.60) = 0.144

b. *NR,R,R* (0.40)(0.60)(0.60) = 0.144

Σ = 0.432

$$P(x)=\frac{n!}{x!(n-x)!}(\pi)^x(1-\pi)^{n-x}=\frac{3!}{2!(3-2)!}(0.60)^2(0.40)^1=0.432$$

Self-Review 6.3

Use Appendix A and $n=12$

a. $P(0)=0.014$ b. $P(x\geq 5)=0.158+0.079+0.029+0.008+0.001+0+0+0=0.275$

c. $P(2\leq x\leq 4)=0.168+0.240+0.231=0.639$

Self-Review 6.4

$\mu=np=(1000)(0.002)=2.00$ and Appendix C, $\mu=2$ and $x=0$

a. $P(0)=0.1353$ or by formula $P(x)=\frac{u^x e^{-u}}{x!}=\frac{2^0 e^{-2}}{0!}=e^{-2}=0.1353$

b. $P(x\geq 2)=\{1-[P(0)+P(1)]\}=\{1-[0.1353+0.2707]\}=\{1-0.406\}=0.594$

CONTINUOUS PROBABILITY DISTRIBUTION

Self-Review 7.1

a. Determine the height using Formula [7 – 3]

$$P(x) = \frac{1}{(b-a)} = \frac{1}{(3-0)} = \frac{1}{3} = 0.333$$

The uniform distribution is shown.

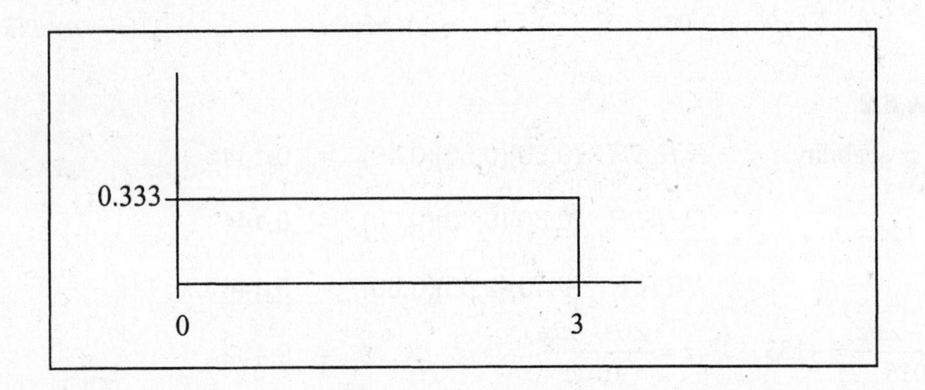

b. The mean represents the typical wait time. To determine the mean use Formula [7–1].

$$\mu = \frac{a+b}{2} = \frac{0+3}{2} = \frac{3}{2} = 1.5$$

c. To determine the standard deviation use Formula [7–2].

$$\sigma = \sqrt{\frac{(b-a)^2}{12}} = \sqrt{\frac{(3-0)^2}{12}} = \sqrt{\frac{(3)^2}{12}} = \sqrt{0.75} = 0.866$$

d. The probability a particular customer will wait less than 1 minutes is found by finding the area of the rectangle with a height of 0..333 and a base of (1 – 0).

$$P(0 < wait\,time < 1) = Height \times Base = \frac{1}{(3-0)} \times (1-0) = 0.333 \times 1 = 0.333$$

e. The probability a particular customer will wait between 1.5 and 2 minutes is found by finding the area of the rectangle with a height of 0.333 and a base of (2 – 1.5).

$$P(1.5 < wait\,time < 2) = Height \times Base = 0.333 \times (2-1.5) = 0.333 \times 0.5 = 0.1665$$

This probability is illustrated by the following graph.

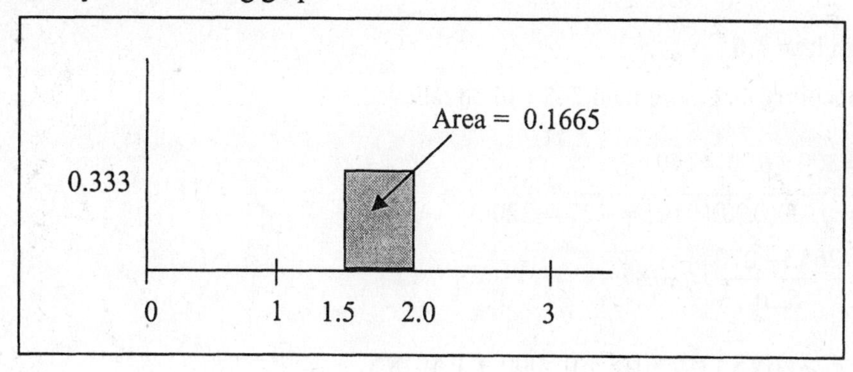

Self-Review 7.2

a. Probability a bottle will contain between 2.02 and 2.04 liters: $z = \frac{X - \mu}{\sigma} = \frac{2.04 - 2.02}{0.015} = 1.33$

Appendix D value is 0.4082; thus the probability a bottle will contain between 2.02 and 2.04 liters is 0.4082.

b. Probability a bottle will contain between 2.00 and 2.03 liters:

$$z = \frac{2.00 - 2.02}{0.015} = -1.33 \qquad z = \frac{2.03 - 2.02}{0.015} = 0.67$$

Appendix D values are 0.4082 and 0.2486.

The two probabilities are added (0.4082 +0.2486 =0.6568), thus the probability of a bottle containing between 2.00 and 2.03 liters is 0.6568.

Self-Review 7.3

Probability a bottle will contain less than 2.00 liters

$$z = \frac{2.00 - 2.02}{0.015} = -1.33$$

Appendix D value is 0.4082, thus (0.5000 – 0.4082 = 0.0918). The probability a bottle will contain less than 2 liters is 0.0918.

Self-Review 7.4

Cola dispensed in the largest 4%. First subtract (0.50 – 0.04) = 0.46, then find the z value such that 0.4600 of the area is between 0 and z. That value is $z = 1.75$. Then solve for X.

$$z = \frac{X - \mu}{\sigma} \qquad 1.75 = \frac{X - 2.02}{0.015} \qquad X = 0.02625 + 2.02 = 2.05 \text{ liters}$$

Thus the largest 4% of the bottles contain 2.05 liters or more.

Self-Review 7.5

The probability that more than 265 will be relieved:

$$\mu = 300 \times 0.90 = 270$$

$$\sigma = \sqrt{300(0.90)(0.10)} = \sqrt{27} = 5.20$$

$$z = \frac{265.5 - 270}{5.20} = -0.87$$

$$P(X > 265.5) = 0.3078 + 0.5000 = 0.8078$$

CHAPTER 8

SAMPLING METHODS AND THE CENTRAL LIMIT THEOREM

Self-Review 8-1

a. Start with row six and column six and use the first two digits .The number is 84822. The column is repeated below with the selected numbers in bold, italics and underlined:
84, 65,32, 23, 36,66, ***17,*** ***16,*** 72,77,***05,*** 27, 35, 33, 21, 51, 28, 72, 71, 90, 69, ***00,*** ***18.***
We select advertiser number ***00, 05, 16, 17, 18.***

b. Starting at the bottom right corner of the table. The number is 70603. The column is repeated below starting with 03 and working up with the selected numbers in bold italics and underlined:
03, 89, 95, ***10,*** 43, ***00,*** ***09,*** 26, 98, 87, 48, 49, ***18.*** We select advertisers: ***00, 03, 09, 10, 18.***

c. Every fourth advertiser starting with 03 would be advertiser: ***03, 07, 11, 15, 19.***

d. In order to select a sample of four advertisers so that one of each type of advertiser is included we had to skip a selection when we had already selected an advertiser for that category. Starting at the top of column two. The number is 08182. Use the left two digits starting with 08. The numbers in the column are repeated below with the selected numbers in bold, italics and underlined:
08,(S), 90, 78, 97, 00(S) Skip, ***19,(P),*** 04(S) Skip, 14(S) Skip, 57, 81, 26, 25, 67, 22, 19(P) Skip, 72, 90, 76, 36, 87, 97, 47, 25, ***15(A),*** 24, 86, 53, 73, 41, ***03 (R)*** We select advertisers ***08(S), 19(P), 15(A), 03(R).***

Self-Review 8.2

a. $_5C_3 = \frac{5!}{3!2!} = 10$

b.

Sample Number	Homes Sold	Total Homes Sold	Mean Number of Homes Sold
1	ABC	13	4.33
2	ABD	17	5.67
3	ABE	11	3.67
4	BCD	16	5.33
5	BCE	10	3.33
6	CDE	17	5.67
7	CDA	20	6.67
8	DEA	18	6.00
9	DEB	14	4.67
10	ACE	14	4.67

c.

Mean Sold	Frequency	Probability
3.33	1	0.1
3.67	1	0.1
4.33	1	0.1
4.67	2	0.2
5.33	1	0.1
5.67	2	0.2
6.00	1	0.1
6.67	1	0.1
	10	1.0

d.

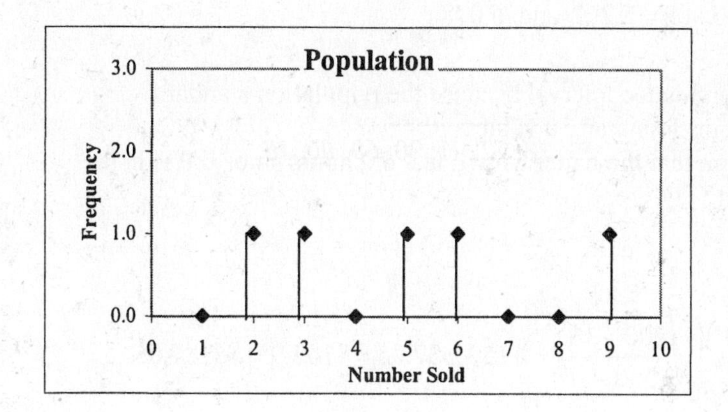

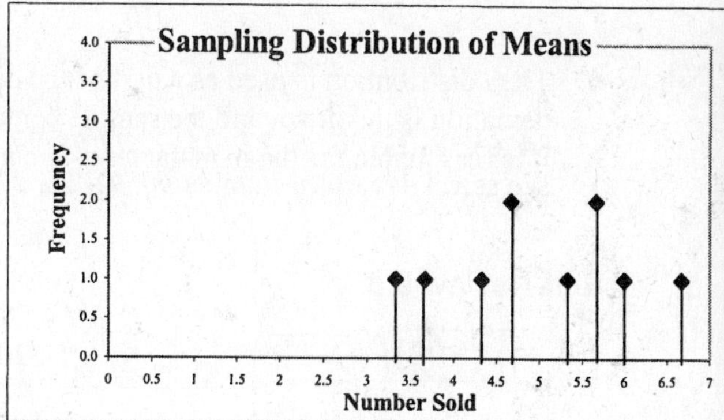

Self-Review 8.3

0.7881, found by: $$z = \frac{\bar{X} - \mu}{\sigma / \sqrt{n}} = \frac{48.45 - 48.5}{0.25/\sqrt{16}} = \frac{-0.05}{0.0625} = -0.8$$

The probability that z is greater than – 0.8 is (0.5000 + 0.2881) = 0.7881

Self-Review 8.4

0.0228, found by: $$z = \frac{\bar{X} - \mu}{s / \sqrt{n}} = \frac{47.88 - 48}{0.48/\sqrt{64}} = \frac{-0.12}{0.06} = -2.00$$

Referring to Appendix D, the z-value for – 2.00 is 0.4772. The likelihood of finding a z-value less than –2.00 is found by (0.5000 – 0.4772) = 0.0228. There is about a 2.28% chance of finding a sample mean of 47.88 ounces or less from the population.

CHAPTER 9

ESTIMATION AND CONFIDENCE INTERVALS

Self-Review 9.1

The z value is 2.33, found by Appendix D and locating the value 0.4901 in the body of the table, and reading the corresponding row and column values.

$$\bar{X} \pm z\frac{s}{\sqrt{n}} = \$150 \pm 2.33\left(\frac{\$20}{\sqrt{36}}\right) = \$150 \pm \$7.77 = \$142.23 \text{ to } \$157.77$$

Self-Review 9.2

a. The sample mean is given: $\bar{X} = 5.67$. The standard deviation is given: $s = 0.57$.

b. The population mean is not known. The best estimate is the sample mean of 5.67.

c. Construct a 95 percent confidence interval for the population mean. Use formula [9-2].

$$\bar{X} \pm t\frac{s}{\sqrt{n}} = 5.67 \pm 2.262\frac{0.57}{\sqrt{10}} = 5.67 \pm 0.408 = 5.262 \text{ and } 6.078$$

d. The t distribution is used as a part of the confidence interval because the population standard deviation is unknown and the sample contains less than 30 values.

e. It is reasonable for the manufacturer to claim that the batteries will last 6.0 hours since 6.0 is in the interval.

Self-Review 9.3

$$p \pm z\sqrt{\frac{p(1-p)}{n}}\left(\sqrt{\frac{N-n}{N-1}}\right) = \$150 \pm 2.33\left(\frac{\$20}{\sqrt{36}}\right)\left(\sqrt{\frac{200-36}{200-1}}\right) = \$150 \pm \$7.05 = \$142.95 \text{ to } \$157.05$$

Self-Review 9.4

The z value is 1.65, found by Appendix D and locating the value 0.4500 in the body of the table, and reading the corresponding row and column values. Note that 0.4500 is exactly half way between 0.4495 and 0.4505.

$$p \pm z\sqrt{\frac{p(1-p)}{n}} = 0.60 \pm 1.65\sqrt{\frac{(0.60)(1-0.60)}{100}} = 0.60 \pm 0.08 = 0.52 \text{ and } 0.68$$

Self-Review 9.5

$$n = \left[\frac{zs}{E}\right]^2 = \left[\frac{(2.58)(0.25)}{0.20}\right]^2 = 10.4 = 11 \text{ (Common practice to round up when determining sample size.)}$$

Self-Review 9.6

$$n = p(1-p)\left(\frac{z}{E}\right)^2 = (0.33)(1-0.33)\left(\frac{2.33}{0.04}\right)^2 = 750.2 = 751$$

CHAPTER 10

ONE-SAMPLE TESTS OF HYPOTHESES

Self-Review 10.1

Step 1: *State the null and alternate hypotheses*: The null hypothesis is that there is no increase in the amount of spending. It is \$30 or less. The alternate hypothesis is that the amount of spending is greater than \$30.

$H_0: \mu \leq \$30$
$H_1: \mu > \$30$

Step 2: *Select the level of significance:* It was given as 0.05.

Step 3: *Select the test statistic:* The test static is the standard normal distribution.

$$z = \frac{\$33 - \$30}{\$12 / \sqrt{40}} = 1.58$$

Step 4: *Formulate the decision rule:*

H_0 is rejected if z is greater than 1.65.

Step 5: *Make a decision regarding the null hypothesis, and interpret the results*:

H_0 is not rejected. No increase in the mean amount spent.

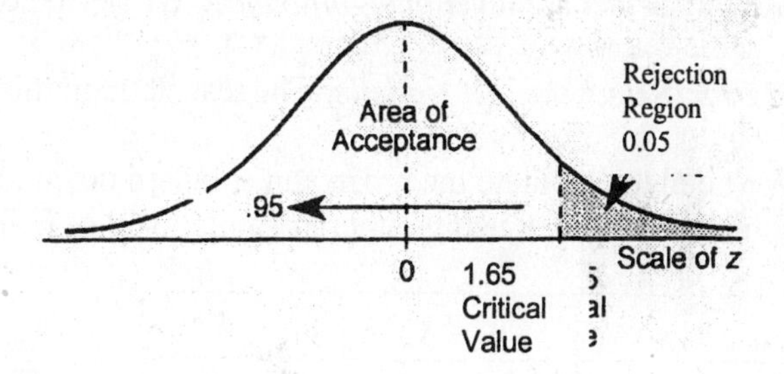

p-value = $P(z > 1.58) = 0.5000 - 0.4429 = 0.0571$
The p-value of $0.0571 \geq 0.05$, do not reject H_0.

Self-Review 10.2

Step 1: *State the null and alternate hypotheses*: The null hypothesis is that there is no change in the construction time. That is, the construction time is at least 3.5 days. The alternate hypothesis is that the construction time is less than 3.5 days. Symbolically, these statements are written as follows:

$H_0: \mu \geq 3.5$
$H_1: \mu < 3.5$

Step 2: *Select the level of significance:* It was given as 0.05.

Step 3: *Select the test statistic:* The test static is the t distribution. The distribution is said to be a normal distribution, however we do not know the value of the population standard deviation. Also, we have a small sample. We use text Formula [9-3]:

Thus: $$t = \frac{\bar{X} - \mu}{s / \sqrt{n}} = \frac{3.4 - 3.5}{0.8 / \sqrt{15}} = \frac{-0.1}{0.20656} = -0.48$$

Step 4: *Formulate the decision rule:* The critical values of t are given in Appendix F. The number of degrees of freedom is $(n - 1) = (15 - 1) = 14$. We have a one-tailed test, so we find the portion of the table labeled 'one-tailed." Locate the column for the 0.05 significance level. Read down the column until it intersects the row with 14 degrees of freedom. The value is 1.761.

Since this is a one-tailed test and the rejection region is in the left tail, the critical value is negative. The decision rule is to reject H_0 if the value of t is less than −1.761.

Step 5: *Make a decision regarding the null hypothesis, and interpret the results*: Because −0.48 lies to the right of the critical value −1.761, the null hypothesis is not rejected at the 0.05 significance level. This indicates that the use of the "precut and assembled roof trusses" does not decrease the construction time to less than 3.5 days.

Self-Review 10.3

Step 1: *State the null and alternate hypotheses*: The null hypothesis is that The University of Findlay students use the computer 2.55 hours a day. The alternate hypothesis is that The University of Findlay students do not use the computer 2.55 hours a day. Symbolically, these statements are written as follows:

$$H_0 : \mu = 2.55$$
$$H_1 : \mu \neq 2.55$$

Step 2: *Select the level of significance:* We decide on the 0.05 significance level.

Step 3: *Select the test statistic:* The test static in this situation is the t distribution.

We need to calculate the mean and standard deviation of the sample. The standard deviation of the sample can be determined using either Formula [3-2] or [3-10].

X	$X - \overline{X}$	$(X - \overline{X})^2$
3.15	0.46154	0.21302
3.25	0.56154	0.31533
2.00	− 0.68846	0.47398
2.50	−0.18846	0.03552
2.65	− 0.03846	0.00148
2.75	0.06154	0.00379
2.35	− 0.33846	0.11456
2.85	0.16154	0.02610
2.95	0.26154	0.06840
2.45	− 0.23846	0.05686
1.95	− 0.73846	0.54532
2.35	− 0.33846	0.11456
3.75	1.06154	1.12687
Σ 34.95	0.00000	3.09579

$$\overline{X} = \frac{\Sigma X}{n} = \frac{34.95}{13} = 2.68846$$

$$s = \sqrt{\frac{\Sigma(X - \overline{X})^2}{n-1}} = \sqrt{\frac{3.09579}{13-1}}$$
$$= \sqrt{0.25798} = 0.5079 = 0.51$$

The value of t is computed using Formula [10-5]:

$$t = \frac{\overline{X} - \mu}{s/\sqrt{n}} = \frac{2.69 - 2.55}{0.51/\sqrt{13}} = \frac{0.14}{0.1414} = 0.9901$$

Step 4: *Formulate the decision rule:*

Remember that the significance level stated in the problem is 0.05. The critical values of t are given in Appendix F. The number of degrees of freedom is $(n - 1) = (13 - 1) = 12$. We have a two-tailed test, so we find the portion of the table labeled "two-tailed." Locate the column for the 0.05 significance level. Read down the column until it intersects the row with 12 degrees of freedom. The value is 2.179.

The decision rule is: Reject the null hypothesis if the computed value of t is to the left of –2.179, or to the right of 2.179

Step 5: *Make a decision regarding the null hypothesis, and interpret the results:*

The value of t lies between the two critical values: –2.179 and 2.179. The null hypothesis is not rejected at the 0.05 significance level. We conclude the population mean hours of usage could be 2.55 hours per day. The evidence fails to show Findlay students to be different.

Self-Review 10.4

$H_0: p \geq 0.40 \qquad H_1: p < 0.40$

The 0.10 level of significance yields a decision rule of (0.500 – 0.100) = 0.400 or 1.28. Thus, H_0 is rejected if z is less than –1.28.

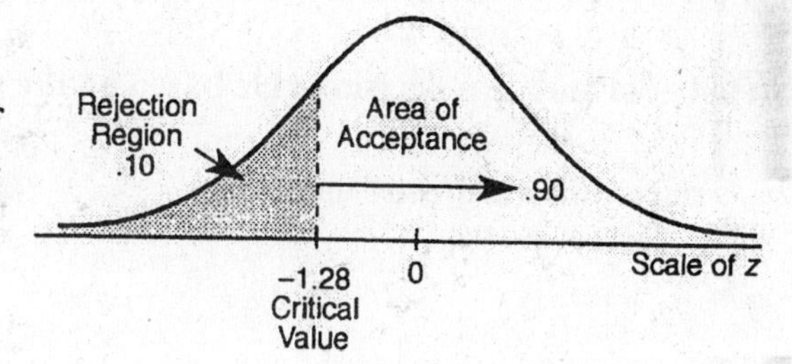

$$z = \frac{\frac{60}{200} - 0.40}{\sqrt{\frac{(0.40)(1-0.40)}{200}}} = \frac{-0.10}{0.03464} = -2.886 = -2.89$$

Since –2.89 lies to the left of –1.28 H_0 is rejected. Less than 40% of the viewing audience watched the concert.

The p-value is the probability of a z value to the left of –2.89. It is (0.5000 – 0.4981) = 0.0019, which is less than the level of significance of 0.10; thus we reject H_0.

CHAPTER 11

TWO-SAMPLE TESTS OF HYPOTHESIS

Self-Review 11.1

Step 1. State the null hypothesis and the alternate hypothesis

Let population 1 refer to Youngsville and population 2 refer to Claredon.

$H_0: \mu_1 \le \mu_2$
$H_1: \mu_1 > \mu_2$

Step 2. Select the level of significance. We have selected 0.05.

Step 3. Identify the test statistic.

The samples are large (greater than 30) the z distribution is used as the test statistic using formula [11-2].

Step 4: Formulate a decision rule based on the selected test statistic and level of significance.

H_0 is rejected if z is greater than 1.65.

$$z = \frac{\bar{X}_1 - \bar{X}_2}{\sqrt{\frac{s_1^2}{n_1} + \frac{s_2^2}{n_2}}} = \frac{6.9 - 4.9}{\sqrt{\frac{(3.8)^2}{60} + \frac{(3.0)^2}{70}}} = 3.29$$

Step 5: Make a decision to reject or not to reject the null hypothesis and interpret the results.

The computed value of z is greater than 1.65. Thus H_0 is rejected. We conclude that it takes Youngsville longer to respond to emergency runs.

Self-Review 11.2

Step 1. State the null hypothesis and the alternate hypothesis

Let population 1 refer to women.

$H_0: \pi_1 \le \pi_2$ $\qquad H_1: \pi_1 > \pi_2$

Step 2. Select the level of significance. We have selected 0.05.

Step 3. Identify the test statistic.

The standard normal distribution is the test statistic using formula [11-3].

Step 4: Formulate a decision rule based on the selected test statistic and level of significance.

H_0 is rejected if z is greater than 1.65.

$$\rho_c == \frac{X_1 + X_2}{n_1 + n_2} = \frac{45 + 25}{150 + 100} = 0.28$$

$$z = \frac{p_1 - p_2}{\sqrt{\frac{p_c(1-p_c)}{n_1} + \frac{p_c(1-p_c)}{n_2}}} = \frac{0.30 - 0.25}{\sqrt{\frac{(0.28)(1-0.28)}{150} + \frac{(0.28)(1-0.28)}{100}}} = 0.86$$

Step 5: Make a decision to reject or not to reject the null hypothesis and interpret the results.

H_0 is not rejected. The proportion of smokers is the same.

p-value = $P(z > 0.86) = (0.5000 - 0.3051) = 0.1949$, The p-value of $0.1949 \geq 0.05$, do not reject H_0.

Self-Review 11.3

Step 1. State the null hypothesis and the alternate hypothesis

Let population 1 refer to the mall and population 2 refer to downtown.

$H_0: \mu_1 \leq \mu_2 \qquad H_1: \mu_1 > \mu_2$

Step 2. Select the level of significance. We have selected 0.01.

Step 3. Identify the test statistic.

The t distribution is the test statistic using formula [11-5 and 11-6].

Step 4: Formulate a decision rule based on the selected test statistic and level of significance.

H_0 is rejected if t is greater than 2.552.

$$s_p^2 = \frac{(n_1 - 1)s_1^2 + (n_2 - 1)s_2^2}{n_1 + n_2 - 2} = \frac{(10-1)(12)^2 + (10-1)(10)^2}{10+10-2} = 122$$

$$t = \frac{\overline{X}_1 - \overline{X}_2}{\sqrt{s_p^2\left(\frac{1}{n_1} + \frac{1}{n_2}\right)}} = \frac{40 - 36}{\sqrt{122\left(\frac{1}{10} + \frac{1}{10}\right)}} = 0.81$$

Step 5: Make a decision to reject or not to reject the null hypothesis and interpret the results.

H_0 is not rejected. There is no difference in mean amount spent at the mall and downtown store. p-value is greater than 0.10.

Self-Review 11.4

Step 1. State the null hypothesis and the alternate hypothesis

$H_0: \mu_d = 0 \qquad H_1: \mu_d \neq 0$

Step 2. Select the level of significance. We have selected 0.05.

Step 3. Identify the test statistic.

The t distribution is the test statistic using formula [11-7].

Step 4: Formulate a decision rule based on the selected test statistic and level of significance.

Reject H_0 if t is less than –2.365 or greater than 2.365.

Electric	Gas	d	$(d-\bar{d})$	$(d-\bar{d})^2$
265	260	5	3.375	11.390625
271	270	1	–0.625	0.390625
260	250	10	8.375	70.140625
250	255	–5	–6.625	43.890625
248	250	–2	–3.625	13.140625
280	275	5	3.375	11.390625
257	260	–3	–4.625	21.390625
262	260	2	0.375	0.140625
	Total	13		171.875

$$\bar{d} = \frac{\Sigma d}{N} = \frac{13}{8} = 1.625$$

$$s_d = \sqrt{\frac{\Sigma(d-\bar{d})^2}{n-1}} = \sqrt{\frac{171.875}{8-1}} = 4.96$$

$$t = \frac{\bar{d}}{s_d/\sqrt{n}} = \frac{1.625}{4.96/\sqrt{8}} = 0.93$$

Step 5: Make a decision to reject or not to reject the null hypothesis and interpret the results.

H_0 is not rejected. There is no difference in the heating cost.

CHAPTER 12

ANALYSIS OF VARIANCE

Self-Review 12.1

$H_0: \sigma_H^2 \le \sigma_T^2$; $\quad H_1: \sigma_H^2 > \sigma_T^2$

At the 0.05 level of significance and 9 degrees of freedom for both the numerator and the denominator, using Appendix G, H_0 is rejected if $F > 3.18$.

$$F = \frac{(60)^2}{(30)^2} = 4.00$$

H_0 is rejected. There is more variation in the Harmon forecast.

Self-Review 12.2

Step 1. State the null hypothesis and the alternate hypothesis

$H_0: \mu_1 = \mu_2 = \mu_3$; $\quad H_1$: Not all means are equal.

Step 2. Select the level of significance. We have selected 0.05.

Step 3. Determine the test statistic.

For an analysis of variance problem the appropriate test statistic is F.

Step 4. Formulate the Decision Rule.

H_0 is rejected if F is greater than 3.59. Degrees of freedom in numerator = (3 – 1) = 2, degrees of freedom in denominator = (20 – 3) = 17

Step 5. Select the sample, perform the calculations, and make a decision.

- Compute the grand mean and the SS total using Table I.

Use columns **A, D, & G**: $\bar{X}_G = \dfrac{\Sigma all\ the\ X\ values}{n} = \dfrac{19+52+40}{5+7+8} = \dfrac{111}{20} = 5.55$

Table I

	A	B	C	D	E	F	G	H	I		
	60 Degrees			70 Degrees			80 Degrees				
	X	$X-\bar{X}_G$	$(X-\bar{X}_G)^2$	X	$X-\bar{X}_G$	$(X-\bar{X}_G)^2$	X	$X-\bar{X}_G$	$(X-\bar{X}_G)^2$	X	
	3	-2.55	6.50	7	1.45	2.10	4	-1.55	2.40		
	5	-0.55	0.30	6	0.45	0.20	6	0.45	0.20		
	4	-1.55	2.40	8	2.45	6.00	5	-0.55	0.30		
	3	-2.55	6.50	9	3.45	11.90	7	1.45	2.10		
	4	-1.55	2.40	6	0.45	0.20	6	0.45	0.20		
				8	2.45	6.00	5	-0.55	0.30		
				8	2.45	6.00	4	-1.55	2.40		
							3	-2.55	6.50		
Σ	19			52			40			111.00	
Σ			18.10			32.40			14.40	64.90	SS Total
n	5			7			8			20	

- Use columns **C, F, & I**: $SS\,total = \Sigma(X-\bar{X}_G)^2 = (18.10+32.40+14.40) = 64.90$

Use Table II to compute the mean for each group.

For 60° $\bar{X} = \dfrac{\Sigma X}{n} = \dfrac{19}{5} = 3.8$ For 70° $\bar{X} = \dfrac{\Sigma X}{n} = \dfrac{52}{7} = 7.43$ For 80° $\bar{X} = \dfrac{\Sigma X}{n} = \dfrac{40}{8} = 5.0$

Table II

	A	B	C	D	E	F	G	H	I	
	60 Degrees			70 Degrees			80 Degrees			
	X	$X-\bar{X}$	$(X-\bar{X})^2$	X	$X-\bar{X}$	$(X-\bar{X})^2$	X	$X-\bar{X}$	$(X-\bar{X})^2$	
	3	-0.80	0.64	7	-0.43	0.18	4	-1	1.00	
	5	1.20	1.44	6	-1.43	2.04	6	1	1.00	
	4	0.20	0.04	8	0.57	0.32	5	0	0.00	
	3	-0.80	0.64	9	1.57	2.46	7	2	4.00	
	4	0.20	0.04	6	-1.43	2.04	6	1	1.00	
				8	0.57	0.32	5	0	0.00	
				8	0.57	0.32	4	-1	1.00	
							3	-2	4.00	
Σ	19			52			40			111.00
Σ			2.80			7.68			12.00	22.48
n	5			7			8			20

Use columns **C, F, & I**: $SSE = \Sigma(X - \bar{X}_c)^2 = (2.80 + 7.68 + 12.00) = 22.48$

$$SST = SS\ total - SSE$$
$$= 64.90 - 22.48$$
$$= 42.42$$

ANOVA Table

Source Variation	**Sum of Squares**	**Degrees of Freedom**	**Mean Squares**
Treatment	SST = 42.42	$(k - 1) = (3 - 1) = 2$	$\text{MST} = \frac{\text{SST}}{(k-1)} = \frac{42.42}{2} = 20.65$
Error	SSE = 22.48	$(n - k) = (20 - 3) = 17$	$\text{MSE} = \frac{\text{SSE}}{(n-k)} = \frac{22.48}{17} = 1.32$
Total	SS Total = 64.90	19	

$F = \frac{20.65}{1.32} = 15.64$ H_0 is rejected since F is greater than 3.59.

There is a difference in the mean number correct (achievement).

Self-Review 12.3

The means differ, found by using formula [12-5] and:

$\bar{X}_1 = 7.4$ $n_1 = 7$
$\bar{X}_2 = 3.8$ $n_2 = 5$

$t = 2.110$ from Appendix F: $(n - k) = (20 - 3) = 17$ degrees of freedom and the 95 percent level of confidence.

MSE = 1.32 from SSE/$(n - k)$ = 22.48/17 = 1.32.

$$(\bar{X}_1 - \bar{X}_2) \pm t\sqrt{\text{MSE}\left(\frac{1}{n_1} + \frac{1}{n_2}\right)}$$

$$(7.4 - 3.8) \pm 2.110\sqrt{1.3241\left(\frac{1}{7} + \frac{1}{5}\right)}$$

$$3.6 \pm 1.42167$$

$$2.1783 \text{ to } 5.0217$$

The means differ since both end points of the confidence interval are of the same sign, positive in this problem.

Self-Review 13.1

a. Scatter diagram:

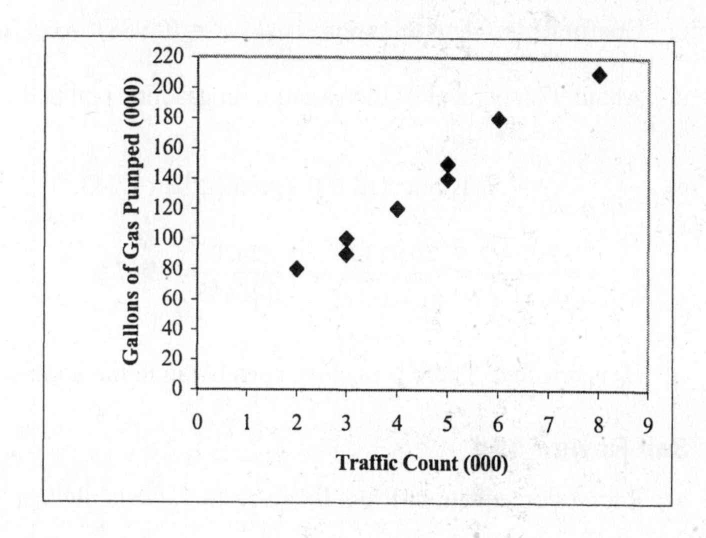

b. Coefficient of correlation:

	A	B	C	D	E	F	G
	Total Gallons			Traffic Count			
	Y	$(Y-\bar{Y})$	$(Y-\bar{Y})^2$	X	$(X-\bar{X})$	$(X-\bar{X})^2$	$(X-\bar{X})(Y-\bar{Y})$
	120	-13.75	189.06	4	-0.5	0.25	6.88
	180	46.25	2,139.06	6	1.5	2.25	69.38
	140	6.25	39.06	5	0.5	0.25	3.13
	150	16.25	264.06	5	0.5	0.25	8.13
	210	76.25	5,814.06	8	3.5	12.25	266.88
	100	-33.75	1,139.06	3	-1.5	2.25	50.63
	90	-43.75	1,914.06	3	-1.5	2.25	65.63
	80	-53.75	2,889.06	2	-2.5	6.25	134.38
n	8			8			
Total	1070		14,387.48	36		26	605.04

Step 1. Compute the means using sums in Column **A** and **D**:

$$\bar{Y} = \frac{\Sigma Y}{n} = \frac{1070}{8} = 133.75 \qquad \bar{X} = \frac{\Sigma X}{n} = \frac{36}{8} = 4.5$$

Step 2. Compute the standard deviations using the sums in Column **C** and **F**:

$$s_y = \sqrt{\frac{\Sigma(Y-\bar{Y})^2}{n-1}} = \sqrt{\frac{14,387.48}{8-1}} = \sqrt{2055.354} = 45.336 = 45.34$$

$$s_x = \sqrt{\frac{\Sigma(X-\bar{X})^2}{n-1}} = \sqrt{\frac{26}{8-1}} = \sqrt{3.714} = 1.927 = 1.93$$

Step 3. Compute the coefficient of correlation ***r*** using the formula, the sum from Column **G** in the table, and the calculated standard deviations.

$$r=\frac{\Sigma(X-\bar{X})(Y-\bar{Y})}{(n-1)(s_x s_y)}=\frac{605.04}{7(45.34)(1.93)}=\frac{605.04}{612.5434}=0.9877=0.988$$

c. Coefficient of determination: $r^2=(0.988)^2=0.976.$

d. About 97.6 percent of the variation in gasoline pumped is explained by the traffic count.

e. $H_0:\rho\le 0$
$H_1:\rho>0$
H_0 is rejected if t is greater than 1.943.

$$t=\frac{r\sqrt{n-2}}{\sqrt{1-r^2}}=\frac{0.988\sqrt{6}}{\sqrt{1-(0.988)^2}}=\frac{2.420}{0.15445}=15.67$$

H_0 is rejected. There is positive correlation in the population.

Self-Review 13.2

a. Regression equation: From Exercise 13.1, we know that: $r=0.988$, $s_y=45.34$, $s_x=1.93$

$$b=r\left(\frac{s_y}{s_x}\right)=0.988\left(\frac{45.34}{1.93}\right)=0.988(23.49)=23.208=23.21$$

In Exercise 13.1, we computed: $\bar{Y}=133.75$ and $\bar{X}=4.5$. Thus using formula [13-5]:

$$\begin{aligned}a&=\bar{Y}-b\bar{X}\\&=133.75-[(23.21)(4.5)]\\&=133.75-104.45=29.30\end{aligned}$$

Using the computed values the regression equation is:

$Y'=a+bX=29.30+23.21\,X$ (in dollars)

b. Standard error of estimate:

X	bX	Y'	Y	$(Y-Y')$	$(Y-Y')^2$	Y^2	XY
4	92.84	122.14	120	-2.140	4.580	14400	480
6	139.26	168.56	180	11.440	130.874	32400	1080
5	116.05	145.35	140	-5.350	28.622	19600	700
5	116.05	145.35	150	4.650	21.623	22500	750
8	185.68	214.98	210	-4.980	24.800	44100	1680
3	69.63	98.93	100	1.070	1.145	10000	300
3	69.63	98.93	90	-8.930	79.745	8100	270
2	46.42	75.72	80	4.280	18.318	6400	160
Sum			1070		309.707	157500	5420

$$s_{y\cdot x} = \sqrt{\frac{\Sigma(Y - Y')^2}{n-2}} = \sqrt{\frac{309.707}{8-2}} = \sqrt{51.6178} = 7.18$$

$$s_{y\cdot x} = \sqrt{\frac{\Sigma Y^2 - a\Sigma Y - b\Sigma XY}{n-2}} = \sqrt{\frac{157,500 - 29.30(1,070) - 23.21(5,420)}{8-2}} = 7.646$$

c. A 95% confidence interval:

$$Y' = 29.30 + 23.21X = 29.30 + 23.21(4) = 122.14$$

$$Y' \pm t(s_{y\cdot x})\sqrt{\frac{1}{n} + \frac{(X - \bar{X})^2}{\Sigma(X - \bar{X})^2}} = 122.14 \pm 2.447(7.646)\sqrt{\frac{1}{8} + \frac{(4-4.5)^2}{26}}$$

$$= 122.14 \pm 6.8646$$

$$= 115.275 \text{ to } 129.005$$

d. A 95% prediction interval:

$$Y' \pm t(s_{y\cdot x})\sqrt{1 + \frac{1}{n} + \frac{(X - \bar{X})^2}{\Sigma(X - \bar{X})^2}} = 122.14 \pm 2.447(7.646)\sqrt{1 + \frac{1}{8} + \frac{(4-4.5)^2}{26}}$$

$$= 122.14 \pm 19.929$$

$$= 102.21 \text{ to } 142.069$$

Self-Review 13.3

a. Coefficient of determination:

$$r^2 = \frac{SSR}{SS\,Total} = \frac{14,078}{14,388} = 0.9785 \quad OR \quad 1 - \frac{SSE}{SS\,Total} = 1 - \frac{310}{14,388} = (1 - 0.0215) = 0.9785$$

b. Coefficient of correlation: $r = \sqrt{r^2} = \sqrt{0.9785} = 0.9892$

c. Standard error of estimate: $s_{y.x} = \sqrt{\frac{SSE}{n-2}} = \sqrt{\frac{310}{8-2}} = 7.188$

CHAPTER 14

MULTIPLE REGRESSION AND CORRELATION ANALYSIS

Self-Review 14.1

No problem with multicollinearity since the correlation of 0.506 between the independent variables temperature and output is in the rule of thumb range of –0.70 to + 0.70.

$H_0: \beta_1 = \beta_2 = 0$ $\qquad$ H_0 : Not all β s are zero

H_0 is rejected if $F > 4.74$, found by Appendix G, $\alpha = 0.05$, $df_n = 2$, $df_d = 7$

$$F = \frac{\frac{17.069}{2}}{\frac{7.031}{7}} = \frac{8.5345}{1.004} = 8.5$$

H_0 is rejected. At least one regression coefficient is not equal to zero.

Self-Review 14.2

For temperature:

$H_0: \beta_1 = 0$

$H_1: \beta_1 \neq 0$

For Output

$H_0: \beta_2 = 0$

$H_1: \beta_2 \neq 0$

Reject H_o if $t < -2.365$ or $t > 2.365$

For temperature $t = 3.72$, thus H_0 for β_1 is rejected. It appears temperature is related to usage.

For output $t = -0.36$, thus H_0 for β_2 is not rejected. It appears that output is not related to usage.

Self-Review 14.3

a. The regression equation is: $Y' = -17.2 + 0.353 X_1$ where Y' is usage and X_1 is temperature.

b. Y' for the first sample item is found by substituting the value: $X_1 = 83$ into the equation.

$$Y' = -17.2 + 0.353 X_1$$
$$= -17.2 + 0.353(83) = 12.099 = 12.10$$

c. The coefficient of multiple determination is R-Sq on the printout. It is 70.3%. A total of 70.3% of the variation in power usage is explained by the temperature.

d. The MINITAB System was used to develop the fitted values of Y' and the residuals. The residuals should approximate a normal distribution. The residuals were organized into the histogram as shown. The shape seems to be somewhat like the normal distribution.

e. The scatter diagram is used to investigate homoscedasticity. The horizontal axis is the fitted values, i.e., Y', and the vertical axis reflects the residuals. Homoscedasticity requires that the residuals remain constant for all fitted values of Y'. This assumption seems to be met, according to the plot.

CHAPTER 15

CHI-SQUARE APPLICATIONS

Self-Review 15.1.

H_0: Tire failures are uniformly distributed.

H_1: Tire failures are not uniformly distributed.

Reject H_0 if the computed value of χ^2 is greater than 7.815, found by Appendix B, 0.05 level of significance and 3 *df*.

	Col. 1	Col. 2	Col. 3	Col. 4	Col. 5
Location	f_o	f_e	$f_o - f_e$	$(f_o - f_e)^2$	$\frac{(f_o - f_e)^2}{f_e}$
Left Front	28	25	3	9	0.36
Left Rear	20	25	–5	25	1.00
Right Front	29	25	4	16	0.64
Right Rear	23	25	–2	4	0.16
	100	100	0		2.16

H_0 is not rejected. There is no difference in the failure rate.

Self-Review 15.2

H_0: There has been no change in the distribution.

H_1: There has been a change in the distribution..

Reject H_0 if the computed value of χ^2 is greater than 7.815, found by Appendix B, 0.05 level of significance and 3 *df*.

	Col. 1	Col. 2	Col. 3	Col. 4	Col. 5	Col. 6
Company	**Percent of Total**	**Number in Sample** f_o	f_e	$f_o - f_e$	$(f_o - f_e)^2$	$\frac{(f_o - f_e)^2}{f_e}$
GM	42	330	336	–6	36	36/336 = 0.107
Ford	33	275	264	11	121	121/264 = 0.458
DaimlerChrysler	22	174	176	–2	4	4/176 = 0.023
Other	3	21	24	–3	9	9/24 = 0.375
Total	100	800	800	0		$\chi^2 = 0.963$

H_0 is not rejected. There has been no change in the distribution.

Self-Review 15.3

H_0: There is no relationship between gender and amount of time spent watching TV.

H_1: There is a relationship between gender and amount of time spent watching TV.

Reject H_0 if the computed value of χ^2 is greater than 5.991, found by Appendix B, 0.05 level of significance and 2 *df.*

$$f_e = \frac{\text{(row total)(column total)}}{\text{grand total}} = \frac{160 \times 225}{480} = 75, \quad \frac{160 \times 255}{480} = 85$$

	Male			**Female**			
Hours		f_e	$\frac{(f_o - f_e)^2}{f_e}$	f_o	f_e	$\frac{(f_o - f_e)^2}{f_e}$	**Total**
Under 8	70	75	0.333	90	85	0.294	160
8 up to 15	100	75	8.333	60	85	7.353	160
15 or more	55	75	5.333	105	85	4.706	160
Total	225	225	13.999	255	255	12.353	480

$$\chi^2 = \Sigma\left[\frac{(f_o - f_e)^2}{f_e}\right] = \frac{(70-75)^2}{75} + \frac{(100-75)^2}{75} + \ldots + \frac{(105-85)^2}{85} = 26.352$$

H_0 is rejected because χ^2 is greater than 5.991. There is a relationship between gender and the amount of time spent watching television.

APPENDIXES

Tables

Appendix A

Binomial Probability Distribution

Appendix B

Critical Values of Chi-Square

Appendix C

Poisson Distribution

Appendix D

Areas under the Normal Curve

Appendix E

Table of Random Numbers

Appendix F

Student's *t* Distribution

Appendix G

Critical Values of the *F* Distribution

Appendix A

Binomial Probability Distribution (*concluded*)

$n = 14$

Probability

x	0.05	0.10	0.20	0.30	0.40	0.50	0.60	0.70	0.80	0.90	0.95
0	0.488	0.229	0.044	0.007	0.001	0.000	0.000	0.000	0.000	0.000	0.000
1	0.359	0.356	0.154	0.041	0.007	0.001	0.000	0.000	0.000	0.000	0.000
2	0.123	0.257	0.250	0.113	0.032	0.006	0.001	0.000	0.000	0.000	0.000
3	0.026	0.114	0.250	0.194	0.085	0.022	0.003	0.000	0.000	0.000	0.000
4	0.004	0.035	0.172	0.229	0.155	0.061	0.014	0.001	0.000	0.000	0.000
5	0.000	0.008	0.086	0.196	0.207	0.122	0.041	0.007	0.000	0.000	0.000
6	0.000	0.001	0.032	0.126	0.207	0.183	0.092	0.023	0.002	0.000	0.000
7	0.000	0.000	0.009	0.062	0.157	0.209	0.157	0.062	0.009	0.000	0.000
8	0.000	0.000	0.002	0.023	0.092	0.183	0.207	0.126	0.032	0.001	0.000
9	0.000	0.000	0.000	0.007	0.041	0.122	0.207	0.196	0.086	0.008	0.000
10	0.000	0.000	0.000	0.001	0.014	0.061	0.155	0.229	0.172	0.035	0.004
11	0.000	0.000	0.000	0.000	0.003	0.022	0.085	0.194	0.250	0.114	0.026
12	0.000	0.000	0.000	0.000	0.001	0.006	0.032	0.113	0.250	0.257	0.123
13	0.000	0.000	0.000	0.000	0.000	0.001	0.007	0.041	0.154	0.356	0.359
14	0.000	0.000	0.000	0.000	0.000	0.000	0.001	0.007	0.044	0.229	0.488

$n = 15$

Probability

x	0.05	0.10	0.20	0.30	0.40	0.50	0.60	0.70	0.80	0.90	0.95
0	0.463	0.206	0.035	0.005	0.000	0.000	0.000	0.000	0.000	0.000	0.000
1	0.366	0.343	0.132	0.031	0.005	0.000	0.000	0.000	0.000	0.000	0.000
2	0.135	0.267	0.231	0.092	0.022	0.003	0.000	0.000	0.000	0.000	0.000
3	0.031	0.129	0.250	0.170	0.063	0.014	0.002	0.000	0.000	0.000	0.000
4	0.005	0.043	0.188	0.219	0.127	0.042	0.007	0.001	0.000	0.000	0.000
5	0.001	0.010	0.103	0.206	0.186	0.092	0.024	0.003	0.000	0.000	0.000
6	0.000	0.002	0.043	0.147	0.207	0.153	0.061	0.012	0.001	0.000	0.000
7	0.000	0.000	0.014	0.081	0.177	0.196	0.118	0.035	0.003	0.000	0.000
8	0.000	0.000	0.003	0.035	0.118	0.196	0.177	0.081	0.014	0.000	0.000
9	0.000	0.000	0.001	0.012	0.061	0.153	0.207	0.147	0.043	0.002	0.000
10	0.000	0.000	0.000	0.003	0.024	0.092	0.186	0.206	0.103	0.010	0.001
11	0.000	0.000	0.000	0.001	0.007	0.042	0.127	0.219	0.188	0.043	0.005
12	0.000	0.000	0.000	0.000	0.002	0.014	0.063	0.170	0.250	0.129	0.031
13	0.000	0.000	0.000	0.000	0.000	0.003	0.022	0.092	0.231	0.267	0.135
14	0.000	0.000	0.000	0.000	0.000	0.000	0.005	0.031	0.132	0.343	0.366
15	0.000	0.000	0.000	0.000	0.000	0.000	0.000	0.005	0.035	0.206	0.463